George Bond Howes

An atlas of practical elementary biology

George Bond Howes

An atlas of practical elementary biology

ISBN/EAN: 9783337276652

Printed in Europe, USA, Canada, Australia, Japan

Cover: Foto ©berggeist007 / pixelio.de

More available books at **www.hansebooks.com**

AN

ATLAS

OF

PRACTICAL ELEMENTARY

BIOLOGY

BY

G. B. HOWES,

DEMONSTRATOR OF BIOLOGY, NORMAL SCHOOL OF SCIENCE AND ROYAL SCHOOL OF MINES, LECTURER ON COMPARATIVE ANATOMY,
ST. GEORGE'S HOSPITAL MEDICAL SCHOOL, LONDON.

WITH A PREFACE
BY
PROFESSOR HUXLEY, P.R.S.

London:
MACMILLAN AND CO.,
1885.

TO
A. H.,
FRIEND AND ADVISER,
THIS WORK IS DEDICATED
BY
THE AUTHOR.

AUTHOR'S PREFACE.

This work has been designed with a view of furnishing the beginner with an evenly balanced series of drawings, illustrative of the typical facts in the structure of living things. The types chosen are for the most part identical with those adopted in Huxley and Martin's "Elementary Biology."

The information imparted by a competent teacher ought to receive ample illustration at his hands, and while it is hoped that this book may be of service to the student thus happily placed, in producing it the author has been especially mindful of the less fortunate inquirer, compelled to work unaided in a field beset with snares and pitfalls, and byways which lead only to a laborious idleness.

An extensive and fully illustrated literature is within reach of the student, when once he has acquired that method which can alone enable him to use it rightly; and where this is the case for the organisms here dealt with, attempts have been made to supplement it as far as possible.

All the figures are drawn, unless otherwise stated, from preparations made specially with a view to the capacity of this work, and the plates are arranged in that order in which it is most desirable the beginner should work them over. The paper used in printing will take colour, provided the ordinary precautions are observed to avoid going over the same surface twice while wet.

The text is confined exclusively to a description of the precise manner in which each preparation was made, and as a number of valuable papers on many of the subjects dealt with have appeared since the publication of our current zoological textbooks, I append a bibliography of them, together with certain classical monographs indispensable to those desirous of extending either their own knowledge, or that of their fellows, in the matter concerned.

The titles of the above-named works are arranged under heads and numbered for purposes of reference in the text.

In preparing this work, it has been my good fortune to have had the counsel of Professor HUXLEY, and I have to acknowledge my indebtedness to Professors T. J. PARKER and E. R. LANKESTER, and to Mr. F. O. BOWER. My friend Mr. M. M. TERRERO and certain of my pupils have rendered me welcome aid in the matter of cutting sections, and my thanks are due to my lithographer, Mr. M. P. PARKER, for the able manner in which he has carried out my wishes.

GEO. BOND HOWES.

NORMAL SCHOOL OF SCIENCE AND ROYAL SCHOOL OF MINES,
SOUTH KENSINGTON, *February*, 1885.

PREFACE.

When, in the year 1872, the system of practical instruction, which is at present pursued in the Biological Laboratory of the Normal School of Science and Royal School of Mines, was established, one of my first cares was the creation of a teaching collection for the use of the students who were following that course of instruction. This collection was to contain, in the first place, a series of preparations and dissections illustrative of every important fact in the structure of the animals and plants selected for study; and, in the second place, a corresponding series of drawings of the dissections, of large size and executed in such a manner as to facilitate the comprehension of the structures represented.

The construction of such a teaching collection as this has involved the expenditure of a great deal of time and skill; and the whole burden of the work has fallen upon my former demonstrator, Mr. T. J. PARKER (now Professor of Biology at Otago), and Mr. GEORGE HOWES, who succeeded Mr. PARKER, and now holds the office.

Ten years ago, assisted by Prof. MARTIN, I published "A Course of Practical Instruction in Elementary Biology," which exemplifies the method of instruction pursued in the Biological Laboratory, and is intended to take the place of the oral instruction which we supply there. The absence of illustrations, however, has greatly interfered with the usefulness of this work, and I am therefore very glad that Mr. HOWES has undertaken to make good the defect by the publication of the present Atlas, which, while starting from part of the work with which he has been occupied in our Laboratory, contains so many accurate and well-devised additional illustrations that it will be hardly less useful to students who are engaged in the Laboratory than to those who work independently of it.

No doubt, the direct instruction of a teacher is very valuable; but, with the aid of this Atlas, I think that an intelligent student, who is unable to obtain that advantage, will find no difficulty in working through "The Course of Practical Instruction in Elementary Biology" by himself.

T. H. HUXLEY.

South Kensington,
  April 24th, 1885.

CONTENTS.

THE FROG (Plates I. to VII.)	Page 1
THE CRAYFISH (Plates VIII. to X.)	,, 31
THE EARTHWORM (Plates XI., XII.)	,, 45
THE SNAIL (Plates XIII., XIV.)	,, 53
THE MUSSEL (Plates XV., XVI.)	,, 61
THE HYDRA (Plate XVII.)	,, 69
THE UNICELLULAR ORGANISMS (Plate XVIII.)	,, 73
THE FUNGI (Plate XIX.)	,, 77
THE STONEWORTS (Plate XX.)	,, 81
THE FERN (Plates XXI., XXII.)	,, 85
THE FLOWERING PLANT (Plates XXIII., XXIV.)	,, 93
APPENDIX AND BIBLIOGRAPHY	,, 105

THE FROG.

PLATES I. TO VII.

v

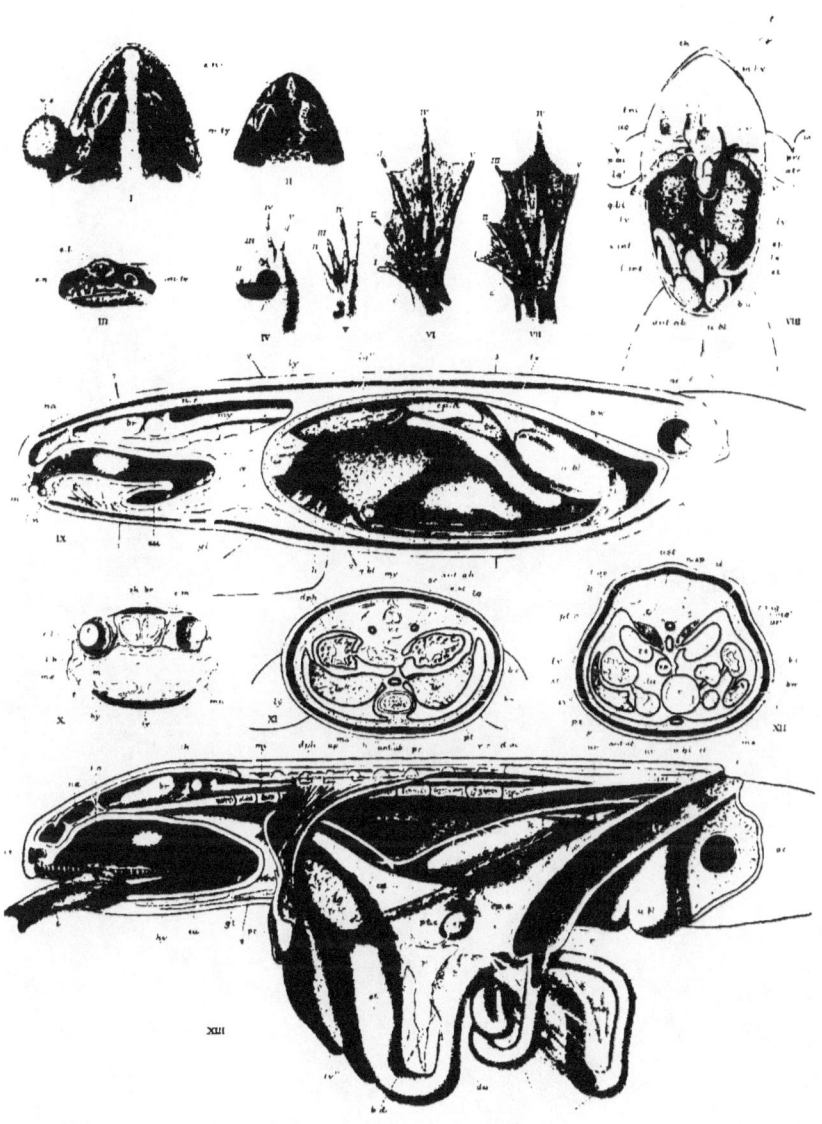

PLATE I.

THE FROG.—EXTERNAL CHARACTERS. THE GREAT CAVITIES OF THE BODY, AND THE GENERAL DISPOSITION OF THE ORGANS CONTAINED THEREIN. THE MODE OF SUSPENSION OF THE VISCERA.

Unless otherwise stated, all the figures of Plates I. to VII. refer to the Common Frog, *Rana temporaria*.

FIG. I.—Head of *Rana esculenta*, ♂, from above.
The vocal sac was inflated from the mouth.

FIG. II.—Head of *Rana temporaria*, ♂, from the same aspect.

FIG. III.—The same, side view.

FIG. IV.—The left manus of *Rana temporaria*.
♂, Palmar surface.

FIG. V.—The same of the ♀.

FIG. VI.—The left pes of *Rana temporaria*. Plantar surface.

FIG. VII.—The same of *Rana esculenta*.

(Figs. I. to VII. all natural size.)

FIG. VIII.—Dissection to show the general disposition of the viscera.
The skin, ventral muscles, and sternum were all removed, and the pericardium opened up.
All the organs were left undisturbed, with the exception of the right lung; in removing the ventral muscles the anterior abdominal vein was dissected out, and the great veins of the head and arm were removed on the right side,* to display the arterial arches and adjacent glands. Nat. size.

FIG. IX.—General dissection, to show the boundaries of the body cavity, and the mutual relations of those organs lodged within it, as seen from the left side (compare Fig. VIII.)
The animal was pinned down through its hind limbs, those of the left side being removed, the pleuro-peritoneal, buccal, nasal, and neural cavities were each opened up, cutting well to one side of the middle line.
The dotted lines indicate the course of the œsophagus. ♂. × 2½.

See Appendix A.

ATLAS OF BIOLOGY.

FIGS. X., XI., XII.—Three sections, taken in order through the planes indicated by the vertical lines in Fig. IX.

Figs. XI. & XII. are especially designed to show the mode of attachment of the viscera, the reflections of the peritoneum being indicated by a thick line.

All are drawn as seen from behind, and the sub-cutaneous lymph-spaces, *ly.*, are indicated in black.   δ.   × 3.

FIG. XIII.—Dissection from the left side, to show the general disposition of the viscera, and the neural canal with the cerebro-spinal axis.*

Proceeding as for Fig. IX., the mesentery was next removed along the line *c.mg'*., and with it the renal and reproductive organs of the left side—those of the right, together with the lung of that side, being seen through the mesentery. The alimentary canal was pinned down and the large intestine opened up; the tongue is figured thrown forward in prehension.

The axial skeleton is here drawn in median longitudinal section. δ. × 2.

* See Appendix B.

ac.	Acetabulum.	*gl.*	Glottis.
ant.ab.	Anterior-abdominal vein.	*h.*	Heart.
ao.	Arch of Aorta.	*hy.*	Hyoid cartilage.
atr.	Atrium.	*i.*	Ileum.
b.c.	Body cavity.	*il.*	Ilium.
b.d.	Bile duct.	*i.n.*	Internal nostril.
br.	Brain.	*k'.*	Right kidney.
b.w.	Body wall.	*k".*	Left kidney.
c.	Calcar.	*l.b.*	Levator bulbi muscle.
ca.	Carotid artery.	*lg'.*	Right lung.
cl.	Cloaca.	*lg".*	Left lung.
cl'.	Epidermal-lined portion of the same.	*l.int.*	Large intestine.
c.mg.	Sub-vertebral lymph sinus.	*lv'.*	Right lobe of liver.
c.mg'.	Its cut edge.	*lv".*	Left lobe of liver.
co.	Rudimentary cæcum.	*ly.*	Sub-cutaneous lymph-spaces.
cm.	Cœlisco-mesenteric artery.	*lx.*	Larynx.
cp.a.	Corpus adiposum.	*m.*	Mouth.
d.ao.	Dorsal aorta.	*m.hy.*	Mylo-hyoid muscle.
dph.	So-called diaphragm.	*mn.*	Mandible.
du.	Duodenum.	*ms.*	Mesentery (dorsal mesentery).
e.	Eye.	*ms'.*	Ventral mesentery.
e.l.	Eye lids.	*m.ty.*	Membrana tympani.
r.m.	Eye muscles.	*mx.*	Maxilla.
e.n.	External nostril.	*my.*	Myelon.
eu.	Eustachian recess.	*na.*	Nasal sac.
g.bl.	Gall bladder.	*n.*	Neural canal.

n.sp.	Spinal nerves.	st.	Stomach.
n.sy.	Sympathetic nerves.	t.	Tongue.
œ.	Œsophagus.	th.	Thyroid gland.
p.	Pancreas.	tm.	Thymus gland.
pc.	Cut edge of pericardium.	ts.	Testis.
p.cu.	Pulmo cutaneous artery.	u.bl.	So-called urinary bladder.
pr.c.	Vena cava superior (precaval vein).	u.bl'.	Orifice of the same.
pt.	Peritoneum.	ur.	Genito-urinary duct ♂.
pt.c.	Vena cava inferior (postcaval vein).	ur'.	Orifice of the same.
r.	Rectum.	ust.	Urostyle.
s.	Sternum.	v.	Ventricle.
s.int.	Small intestine.	v.c.	Vertebral column.
s.sc.	Supra scapula.	v.s.	Vocal sac.
sk.	Skull.	i. to r.	Digits i. to v.
sp.	Spleen.		

PLATE II.

THE FROG.—THE ALIMENTARY, RESPIRATORY, AND URINO-GENITAL ORGANS. THE HEART.
THE LYMPH-HEARTS, ETC.

FIG. I.—The entire alimentary apparatus, after removal from the body, dissected from the ventral aspect.

The lower jaw was pinned back, and the interior of the right vocal sac exposed; the stomach, duodenum, cloaca, and bladder were each opened up. δ. × 2.

The bile-ducts can readily be demonstrated, by gently squeezing the gall-bladder and thus forcing the bile into them. For a full description, with figures, of the smaller hepatic ducts and of the valve-like arrangement of the lining membrane of the small intestine, in *Rana esculenta*, see Wiedersheim and Ecker (24).

FIG. II.—A portion of Fig. I., after removal of the liver. × 1½.

The lymph-sinus drawn can be readily inflated by introducing a blow-pipe under the peritoneum.

FIG. III.—The respiratory segment of Fig. I.
The right lung was opened up, to show its internal structure. × 1½.

FIG. IV.—The left lung of the same, with the larynx seen from the side.

FIG. V.—The same, after removal of the left lung together with one half of the larynx.

FIG. VI.—The cartilages of the larynx dissected out.
Seen from the front. All × 1½.

FIGS. VII. and VIII.—The lymph-hearts. × 1½.

The posterior one, Fig. VIII., is to be seen on removing the skin of the back, immediately after death, and the anterior one by making a clean cut along the ventral body-wall—to one side of the middle line—and a similar one into the cisterna magna, turning the stomach to one side as indicated. Only such parts are drawn as are indispensable in determining the positions of these delicate organs.

Certain details, as to the connections of the above with the great veins, are figured by Ranvier (19).

FIG. IX.—The heart, after removal from the body.
Seen from the front, the aortic arches of the left side having been removed. × 4.

FIG. X.—The same from behind, the sinus venosus having been opened up, to show the sinu-auricular valves. × 4.

FIG. XI.—The same, dissected from the front, the ventral wall together with one of the auriculo-ventricular valves having been removed. × 6.

The rod, passed from the ventricle into the pylangium, indicates the course taken by the blood which flows into the carotid and aortic trunks.

The most recent contribution to the anatomy of the smaller valves of the amphibian heart is that of Boas (2).

FIG. XII.—Dissection of the same, from the left side, all the great cavities opened up. × 4.

From nature, after Huxley's figure for *Rana esculenta* (11).

In preparing the heart from which Figs. IX. to XI. are drawn, the body-cavity of a large frog was laid open, and the whole contents hardened in alcohol* undisturbed. On removing the heart for dissection, the lungs, kidneys, and a portion of the gullet were brought away with it, and finally removed piece by piece, as the great vessels were followed out. By this means a heart, fully distended with its contained blood, may be easily obtained.

FIG. XIII.—The urino-genital organs of the male, dissected from the front, after removal from the body.

The cloaca and bladder were opened up. × 2.

The figure to the left represents the lower portion of the kidney and genito-urinary duct in the ♂ of *Rana esculenta*.

FIG. XIV.—The urino-genital organs of the female, dealt with in the same manner as the above, except that, in order to show the natural relations of the mouth of the oviduct, the left lung and a portion of the oesophagus were also removed from the body.

The genital gland of the right side has been dissected away. × 1½.

* See Appendix C.

ad	Adrenal.	cp.a.	Corpus adiposum.
ao.	Arch of aorta.	du.	Duodenum.
ar.	Arytenoid cartilage.	eu.	Eustachian recess.
a.s.	Inter-auricular septum.	f.t.	Fallopian tube.
au'.	Right auricle.	f.t'.	Mouth of the same.
au".	Left auricle.	g.bl.	Gall-bladder.
b.d.	Common bile-duct.	gl.	Glottis.
b.d'.	Orifice of the same.	h.d.	Hepatic duct.
ca.	Carotid trunk.	hy.	Hyoid bone.
c.d.	Cystic duct.	i.	Ileum.
c.gl.	Carotid gland.	il.	Ilium.
cl.	Cloaca.	i.m.	Intrinsic muscles of larynx.
cl'.	Epidermal-lined portion of the same.	i.n.	Internal nostril.

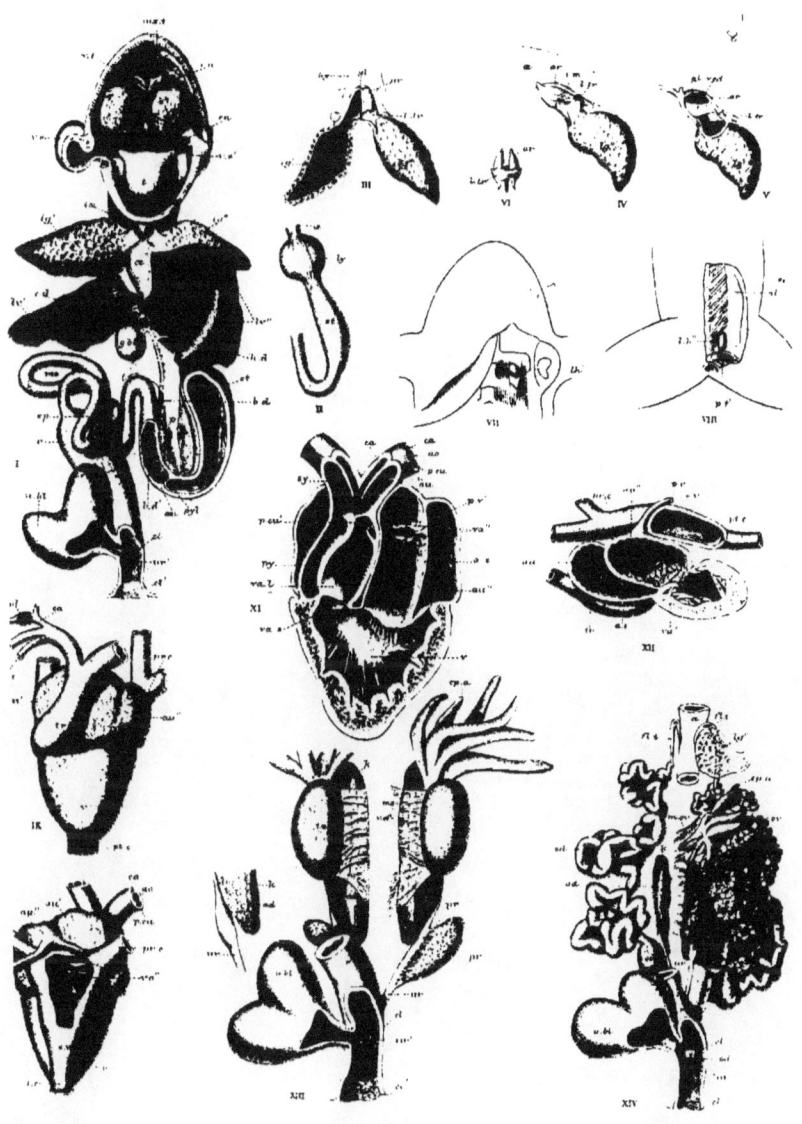

THE FROG.

k.	Kidney.	p.v.	Pulmonary vein.
ly'.	Right lung.	p.v'.	Aperture of entry of the same.
ly".	Left lung.	py.	Pylangium.
l.h'.	Anterior lymph-heart.	pyl.	Pylorus.
l.h".	Posterior lymph-heart.	r.	Rectum.
l.tr.	Laryngo-tracheal cartilage.	sp.	Spleen.
lv'.	Right lobe of liver.	st.	Stomach.
lv".	Left lobe of liver.	s.v.	Sinus venosus.
lx.	Larynx.	sy.	Sypangium.
ly.	Circumœsophageal lymph-sinus.	t.	Tongue.
m.o.	Mesorchium.	tr.	Truncus arteriosus.
m.ov.	Mesoarium.	ts.	Testis.
ms.	Mesentery.	u.bl.	So-called urinary bladder.
mx.t.	Maxillary teeth.	ur.	Ureter ♀. Genito-urinary duct ♂.
od.	Oviduct.	ur'.	Orifice of the same.
od'.	Its uterine portion.	v.	Ventricle.
od".	Aperture of oviduct.	v'.	Cut edge of ventricle.
œ.	Œsophagus.	va'.	Auriculo-ventricular valve.
ov.	Left ovary.	va".	Sinu-auricular valve.
p.	Pancreas.	va.l.	Longitudinal valve (septum) of pylangium.
p.cu.	Pulmo-cutaneous trunk.	va.s.	Semi-lunar valves.
p.cu'.	Point of origin of the same.	v.cd.	Vocal cord.
pf.	Pyriformis muscle.	v.ef.	Efferent ducts of testis.
pr.	So-called vesiculæ-seminales.	v.s.	Vocal sac.
pr'.	Ducts of the same.	v.s'.	Aperture of the left vocal sac.
pv.c.	Vena cava superior.	v.t.	Vomerine teeth.
pv.c.	Vena cava inferior.		

PLATE III

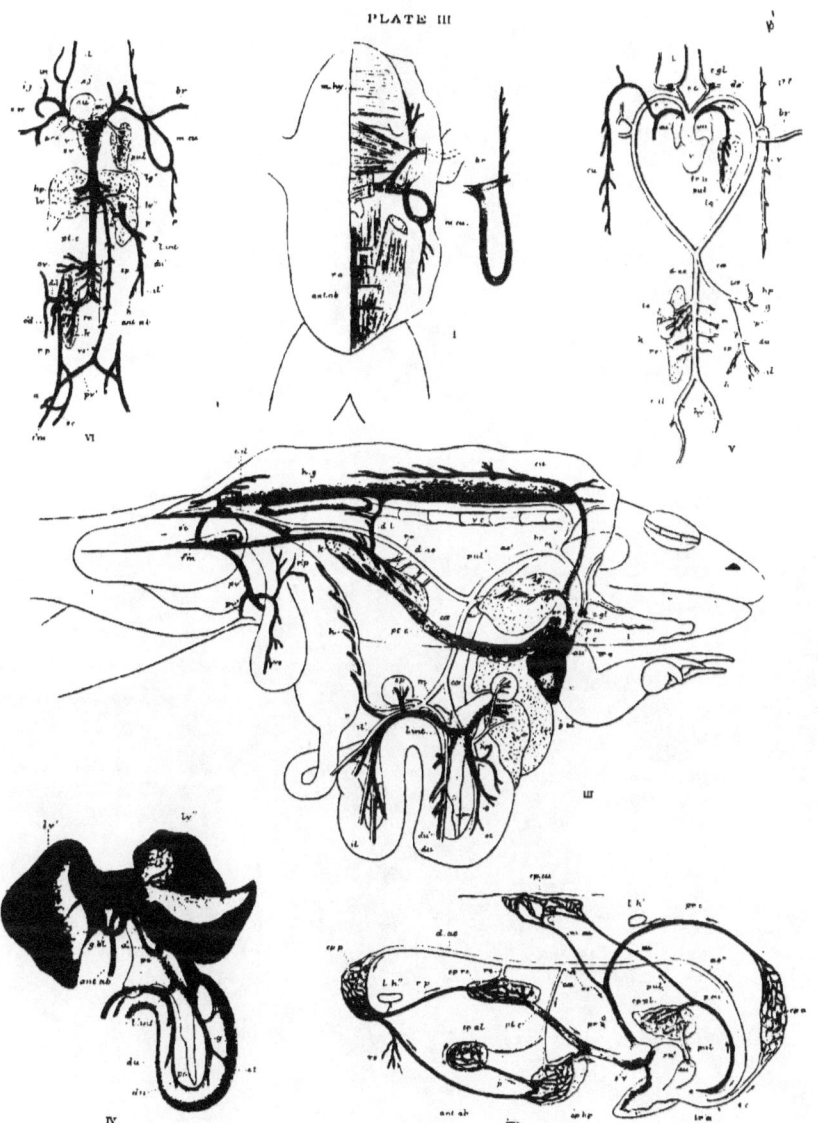

PLATE III.

THE FROG.—The Blood-vascular System.
(If necessary, this system can be readily injected. See Appendix D.)

Fig. I.—A dissection to show the large superficial veins.
The skin covering the ventral surface was reflected on the left side and the pectoral muscles in part removed, as indicated by their cut edges. Nat. size.

Fig. II.—The musculo-cutaneous vein in *Rana esculenta*. Nat. size.

Fig. III.—General dissection of the blood-vessels from the right side.
The body-wall and arm of the right side were cut away, and the skin reflected. After removal of the generative organs, the heart and liver were turned forwards, and sufficient of the latter removed to show the whole course of the vena cava inferior and the hepatic veins. The hind limb of the right side was retained, and the pelvic girdle *h.g.* removed, between the cut ends indicated. This being done, the great veins of the head were dissected away, in order to expose the aortic arches. × 2.

Fig. IV.—Dissection to show the anastomosis between the systemic (anterior-abdominal) and portal venous systems.
The parts are all drawn as they lay, the heart and hinder portion of the left lobe of the liver *lv."* having been turned forwards. × 1½.

Fig. V.—The arterial trunks, and their main branches. × 1½.

Fig. VI.—The venous trunks, and their main factors. × 1½. See Appendix E.

Fig. VII.—A somewhat diagrammatic rendering of Fig. III., representing the entire vascular system as seen from the left side, with all the great capillary systems indicated.
The arrows follow the course taken by the blood in life. × 2.

In all the figures of this plate, the vessels connected with the renal-portal system are shaded darkly, the anterior abdominal vein included, while those carrying arterial blood are left white.

a.	Femoro-sciatic anastomosis.	*au".*	Left auricle.
a'.	Abdominal-portal anastomosis.	*br.*	Brachial vessel.*
ant.ab.	Anterior-abdominal vein.	*c.c.*	Common-carotid artery.
ao'.	Right aortic arch.	*c.gl.*	Carotid gland.
ao".	Left aortic arch.	*c.il.*	Common-iliac artery.
au'.	Right auricle.	*cœ.*	Cœliaco-mesenteric artery.

The references marked thus * apply to both arteries and veins, as the case may be.

ATLAS OF BIOLOGY.

cœ'.	Cœliac artery.	*lr".*	Left lobe of liver.
cp.a.	Capillaries of anterior extremities.	*m.cu.*	Musculo-cutaneous vein.
cp.al.	Capillaries of alimentary canal.	*m.hy.*	Mylo-hyoid muscle.
cp.cu.	Cutaneous capillaries.	*ms.*	Mesenteric artery.
cp.hp.	Hepatic capillaries.	*od.*	Oviducal vein.
cp.p.	Capillaries of posterior extremities.	*ov.*	Ovarian vein.
cp.pl.	Pulmonary capillaries.	*p.*	Portal vein.
cp.re.	Renal capillaries.	*pc.*	Pancreas.
cu.	Cutaneous artery.	*pc'.*	Pancreatic artery.
d.ao.	Dorsal aorta.	*p.cu.*	Pulmo-cutaneous artery.
d.l.	Dorso-lumbar vein.	*pr.c.*	Vena cava superior (Precaval vein).
du.	Duodenum.	*pt.c.*	Vena cava inferior (Postcaval vein).
du'.	Duodenal vessel.*	*pul.*	Pulmonary vein.
e.j.	External jugular vein.	*pul'.*	Pulmonary artery.
fm.	Femoral vein.	*pv'.*	Right pelvic vein.
g.	Gastric vessel.*	*pv".*	Left pelvic vein.
g.bl.	Gall-bladder.	*r.*	Rectum.
h.	Hæmorrhoidal vessel.*	*r.a.*	Rectus-abdominis muscle.
h.g.	Pelvic girdle.	*re.*	Renal vessels.*
hp.	Hepatic vessels.*	*r.p.*	Renal-portal vein.
hy.	Hypogastric artery.	*sc.*	Sciatic vein.
i.j.	Internal jugular vein.	*sp.*	Splenic vessels.*
il.	Ileum.	*s.sc.*	Subscapular vein.
il'.	Ileal vessels.*	*st.*	Stomach.
in.	Innominate vein.	*s.v.*	Sinus venosus.
k.	Right kidney.	*tr.a*	Truncus arteriosus.
l.	Lingual vessel.*	*ts.*	Spermatic artery.
lg'.	Right lung.	*v.*	Ventricle.
lg".	Left lung.	*v.c.*	Vertebral column.
l.h'.	Anterior lymph-heart.	*vs.*	Vesical vein.
l.h".	Posterior lymph-heart.	*vt.*	Vertebral artery (printed as *v.* alone Figs. III. & V.)
l.int.	Lieno-intestinal vein.		
lv'.	Right lobe of liver.		

The references marked thus * apply to both arteries and veins, as the case may be.

PLATE IV.

THE FROG.—THE SKELETON. THE MUSCLES OF THE HIND-LIMB.

FIG. I.—The entire axial skeleton, with the limb-girdles and the limbs of the right side, drawn from above with the body in the resting attitude.*
Macerated preparation. × 1½. The humerus and the metacarpal of the second digit in the manus, drawn separately, are those of the male.

FIG. II.—The first vertebra, seen from the front.

FIG. III.—The fourth vertebra, seen from the same aspect.

FIG. IV.—The fifth and sixth vertebræ, from the side, in the natural position.

FIG. V.—The ninth (sacral) vertebra, seen from behind.
The articulations of the centra are shown in section in Fig. XIII., Plate I.

FIG. VI.—The shoulder-girdle, seen from the front, the left scapula having been straightened out.

FIG. VII.—The left half of the hip-girdle, from the side.†

(Figs. II. to VII., all × 1½.)

FIG. VIII.—The skull (cranio-facial apparatus), after removal of the membrane bones of the right side, seen from above.
The mandible and hyoid are excluded. Wet preparation.‡

FIG. IX.—The same, seen from beneath.
In both this and the above, the membrane bones removed are drawn independently; in this figure, however, one half of the parasphenoid is indicated in situ.

FIG. X.—The same, seen from the side, with both mandible and hyoid in place.

FIG. XI.—Back view of Fig. VIII.

FIG. XII.—Sectional view of the above, exclusive of the hyoid. Meckel's cartilage has been exposed.

(Figs. VIII. to XII., all × 2½.)

FIG. XIII.—The body of the hyoid, of a young frog. × 2.
(In all the above figures the cartilaginous parts are stippled.)

---

* The so-called *calcar, c.*, is generally composed of two or more pieces, and represents the remnant of a *sixth digit.* Compare Born, "Die sechste Zehe der Anuren," *Morph. Jahrb.*, vol. i., 1875; Wiedersheim, "Lehrbuch der vergleichenden Anatomie," Jena, 1883. See also Marsh, "The Limbs of Sauranodon," *American Journal of Science and Arts*, vol. xix., 1880.

† The upper cartilaginous articular-end of the ilium may be appropriately termed the *supra-ilium*.

‡ See Appendix F.

FIG. XIV.—The muscles of the hind-limb, ventral aspect.

The superficial muscles are shown on the right side; the sartorius was reflected, the adductor-longus, adductor-magnus, and gastrocnemius were slightly displaced. On the left side the sartorius was left uninjured; it and the adductor-magnus were a little displaced and the rectus-internus reflected, in order to render visible the semi-tendinosus.

FIG. XV.—The muscles of the hind-limb, dorsal aspect.

On the left side, the vastus externus, and on the right the biceps-femoris have been reflected. The semi-membranosus and the head of the gastrocnemius of the right side were displaced.

FIG. XVI.—The deep muscles inserted into the head of the femur. Right side from without.

The cut ends of the muscles removed are all shown.

(Figs. XIV. to XVI. × $1\frac{1}{2}$. In all the bones are stippled.)

ac.	Acetabulum.		f.t.	Flexor-tarsi anterius.
ad.	Adductor magnus.		g.	Glenoid cavity.
ad'.	Adductor longus.		gl.	Gluteus.
ad".	Adductor brevis.		gs.	Gastrocnemius.
ar.	Splenial.		gs'.	Outer head of the same.
as.	Tibiale.		h.	Humerus.
au.	Periotic capsule.		h.g.	Hip-girdle.
b.f.	Biceps femoris.		hy.	Body of the hyoid.
c.	Calcar.		hy'.	Its anterior cornu.
c.a.	Columella auris.		hy".	Its posterior cornu (thyro-hyoid).
ca.	Fibulare.		il.	Ilium.
cd.	Occipital condyle.		i.ps.	Ileo-psoas.
cd'.	Mandibular condyle.		in.f.	Intervertebral foramen.
cl.	Clavicle.		is.	Ischium.
cn.	Vertebral centrum.		lb.	Labial cartilages.
co.	Coracoid.		m.cp.	Metacarpus.
cp.	Carpus.		md.c.	Medial crest.
cr.	Os cruris.		mk.	Meckel's cartilage.
d.	Dentary.		m.mk.	Mento-meckelian bone.
dl.c.	Deltoid crest.		m.ts.	Metatarsus.
e.c.	Extensor cruris brevis.		mx.	Maxilla.
e.o.	Exoccipital.		na.	Nasal.
f.	Femur.		n.a.	Neural arch.
fb.	Fibula.		n.c.	Neural canal.
fo.a.	Anterior fontanelle.		n.p.	Pre-nasal process.
fo.p.	Posterior fontanelle.		n.p'.	Ali-nasal process.
f.ov.	Fenestra ovalis.		n.sp.	Neural spine.
f.pa.	Fronto-parietal.		ob.	Obturatorius.

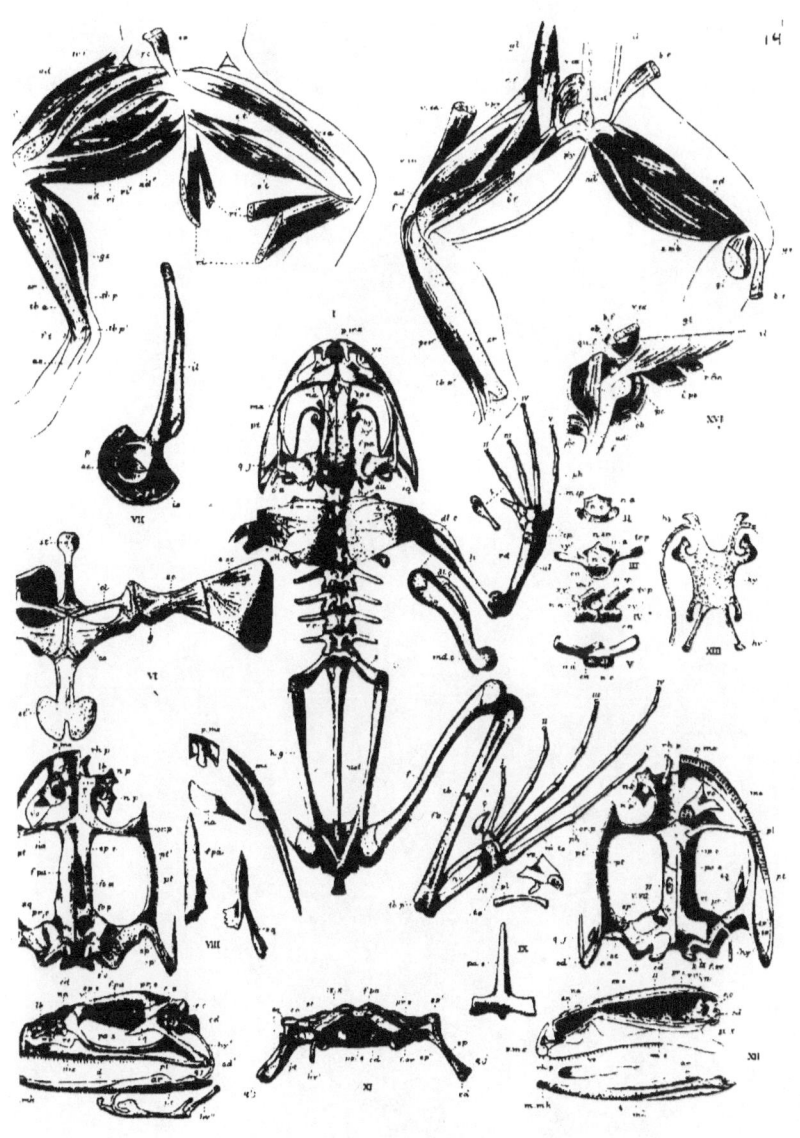

THE FROG.

or.p.	Antorbital process.	*sp'.*	Its dorsal crus.
p.	Pubis.	*sp".*	Its pedicle.
pa.s.	Parasphenoid.	*sp.e.*	Sphenethmoid.
pc.	Pectineus.	*sq.*	Squamosal.
per.	Peroneus.	*st.*	Stapes.
ph.	Phalanges.	*st'.*	Omosternum.
pl.	Palatine.	*st".*	Xiphisternum.
p.mx.	Pre-maxilla.	*s.t.*	Semitendinosus.
pr.o.	Pro-otic.	*s.t'.*	Its union with the adductor magnus.
pt.	Pterygoid bone.	*tb.*	Tibia.
pt'.	Pterygoid cartilage.	*tb.a.*	Tibialis anticus.
py.	Pyriformis.	*tb.p.*	Tibialis posticus.
q.j.	Quadrato-jugal.	*tb.p'.*	Notch for the tendon of the above.
qu.f.	Quadratus femoris.	*tr.f.*	Triceps femoris.
rd.	Radius.	*tr.p.*	Transverse process.
r.f.	Rectus femoris.	*ts.*	Tarsus.
rh.p.	Rhinal process.	*ul.*	Ulna.
r.i.	Rectus internus major.	*ust.*	Urostyle.
r.i'.	Rectus internus minor.	*v.c.*	Vertebral column.
s.	Sacrum.	*v.ex.*	Vastus externus.
sa.	Sartorius.	*v.in.*	Vastus internus.
sc.	Scapula.	*vo.*	Vomer.
s.sc.	Supra scapula.	*zy'.*	Anterior zygapophysis.
sh.g.	Shoulder-girdle.	*zy".*	Posterior zygapophysis.
s.mb.	Semi-membranosus.	*i.* to *v.*	Digits i. to v.
s.n.	Septum nasi.	*I.* to *X.*	Foramina of exit for cranial nerves I. to X.
sp.	Suspensorium.		

PLATE V.

THE FROG.—The Nervous System and Brain, the Anatomy of the Olfactory, Visual, and Auditory Organs.

Fig. I.—The cerebro-spinal axis and the course of the fifth cranial nerve, displayed from above.

The whole dorsal integument was removed, and the neural cavity opened up from above; on the right side the nasal bone has been removed, and the eye dissected out.   × 2.

Fig. II.—The cerebro-spinal axis, with the great nerve trunks—including the sympathetic—dissected from the ventral aspect. The origins of all the cranial nerves except the iv[*] are shown, and the sympathetic is indicated on one side for its whole length.

The entire ventral body-wall, together with the viscera and the floor of the mouth, were removed. After carefully dissecting the aorta away from the sympathetic, that system was removed on the left side. The whole neural canal was next laid open, by removal of the vertebral centra and the floor of the skull; the eye on the left side, the hinder portion of the upper jaw on the right, and the floor of the auditory capsules of both sides, were dissected away. The vii cranial, i spinal, and sympathetic nerves are all shaded darkly.   × 2.

Fig. III.—The cranial nerves, irrespective of those connected with the olfactory, visual, and auditory organs, together with the first five spinal nerves, and sufficient of the sympathetic to show the origin of the splanchnic nerve.

The whole integument covering the head was removed, together with the eye and hinder portion of the upper jaw with its associated parts; the left arm and shoulder-girdle were also removed. The body-cavity and the cisterna-magna were both opened up and the spinal nerves ii to v turned forwards, to render clear the relations of the sympathetic. The vii cranial and sympathetic nerves are tinted as in Fig. II., and all the spinal nerves are drawn in black.   × 2.

Fig. IV.—The leading nerves of the hind limb, seen from the dorsal aspect.

The gastrocnemius, peroneus, and tibialis-anticus muscles have each been reflected, and those of the upper segment of the limb pulled apart, sufficiently to display the nerves drawn. Nat. size.

Fig. V.—A transverse section, taken immediately behind the exit of the second spinal nerve, to show the relations of the nerve roots and investing membranes.   × 6.

Fig. VI.—The brain, seen from above in situ, before the removal of the pia mater.

[*] This is shown in Fig. VIII.

FIG. VII.—Ventral aspect of the same, after removal from the body.*

FIG. VIII.—The same, seen from the left side.
The origin of the fourth cranial nerve is shown.

FIG. IX.—A dissection of the same from above, to show the ventricles.
The lateral and optic ventricles have been exposed on the right side, and the median portion of the cerebellum removed. On the left side the whole system of cavities has been laid bare to the level of the foramen of Monro, in order to show the relationships existing between the axial and lateral portions of the brain.

FIG. X.—The lateral and optic ventricles, exposed from the left side.

FIG. XI.—Vertical longitudinal section of the entire brain, cut a little to one side of the middle line.

Figs. VI. to XI. were drawn from brains preserved in alcohol. All × 3.
The stippled surfaces represent cut edges.

It is doubtful if the parts lettered $ol.$ really represent the olfactory lobes, and the morphology of the nervous tract connecting them has yet to be ascertained. It is also a question if the body here termed pineal gland, $p.gl.$, is the strict homologue of that of other vertebrates. The latest researches in this direction are those of Osborne (17A).

FIG. XII.—The right olfactory sac, opened up from without, along the line a—b of Fig. XIII.
The elevation in its floor overlies the vomer, and the forward growth from the lower lip of the anterior nostril is supported by the alinasal cartilage. ($n.p.'$ Fig. VIII., Pl. IV.) × 4.

FIGS. XIII., XIV., XV.—Three figures of the eye-muscles and their nerves, the nictitating membrane, and Harderian gland. All × 2½.

XIII.—Dorsal view.
The whole integument covering the head was removed, and the superior-rectus of the left side reflected.

XIV.—Ventral view.
The mucous membrane of the roof of the mouth, and the levator-bulbi muscle of the right side, were removed. A portion of the auditory capsule has also been dissected away, to expose the 6th nerve.

XV.—The same; the inferior-oblique and recti muscles, and, on the left side, the outer portion of the face, have all been removed.

FIG. XVI.—The eye, seen from without, in the living animal.

FIG. XVII.—Section of the left eye and eyelids, taken through the plane of the optic nerve.

* The pituitary body is not unfrequently produced into a pair of lateral expansions, thus becoming trilobed.

PLATE V

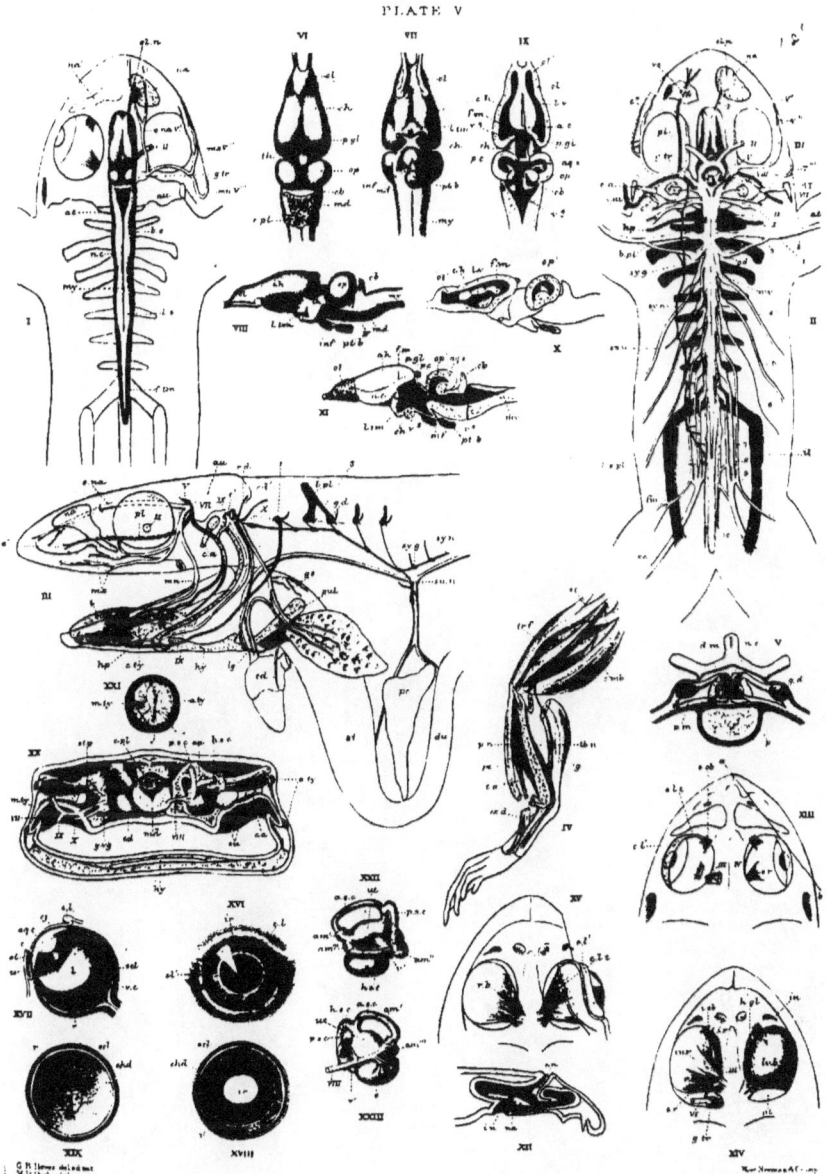

FIG. XVIII.—The outer half of an eye, bisected at right angles to the plane of the optic nerve.

FIG. XIX.—The inner half of the same.
In both figures the internal surface is drawn. All × $3\frac{1}{2}$.

FIG. XX.—Transverse section of the head, immediately behind the occiput, especially designed to show the relations of the auditory organ.
On the left side all structures, excluding the post auditory cranial nerves, have been removed to the level of the columella auris and hind wall of the skull; on the right the auditory capsule has been opened up to the level of the viiith cranial nerve.   × 3.

FIG. XXI.—The membrana tympani, seen from without.
The dotted lines indicate the course of the mucous membrane lining the Eustachian recess, and in both this and the preceding figure, the tympanic annulus is indicated in black.   × 3.

FIG. XXII.—The membranous labyrinth of the left side, seen from without, after removal from the body.
Holding the auditory capsule between the finger and thumb, the bone was sliced away piecemeal with a small scalpel.   × 6.

FIG. XXIII.—The same, seen from within, a portion of the auditory nerve having been removed with it.   × 6.

The structural details of this organ have recently been fully worked out by Retzius, in his classical monograph, "Das Gehörorgan der Wirbelthiere," vol. i., Stockholm, 1881. Figures reproduced in Wiedersheim and Ecker (24).

a.c.	Anterior commissure.	c*.	Commissure between palatine and maxillary nerves.
am'.	Ampulla of the anterior semicircular canal.		
am".	Ampulla of the posterior semicircular canal.	c.a.	Columella auris.
am'''.	Ampulla of the horizontal semicircular canal.	cb.	Cerebellum.
		cd.	Occipital condyle (fig. III. cardiac nerve).
aq.c.	Anterior (aqueous) chamber.	ch.	Optic chiasma.
aq.s.	Aqueduct of Sylvius (Iter).	c.h.	Cerebral hemisphere.
a.s.c.	Anterior-vertical semicircular canal.	chd.	Choroid.
at.	First vertebra.	cj.	Conjunctiva.
a.ty.	Annulus tympanicus.	c.pl.	Choroid plexus.
au.	Auditory capsule.	c.ty.	Chorda tympani.
b.pl.	Brachial plexus.	d.m.	Dura mater.
b.s.	Brachial enlargement.	du.	Duodenum.
c.	Cornea.	e.l.	Upper eyelid.
c'.	Commissure between 7th and 9th cranial nerves.	el'.	Lower eyelid.
		e.l.t.	Tendon of the same.

e.n.	External nostril.	op'.	Ventricle of the same.
e.r.	External rectus.	p.	Periosteum.
eu.	Eustachian recess.	pc.	Pancreas.
e.x.d.	Extensor digitorum muscle.	p.c.	Posterior commissure.
f.m.	Crural nerve.	pe.	Peroneus muscle.
f.m.	Foramen of Monro.	p.gl.	"Pineal gland."
f.tm.	Filum terminale.	pl.	Palatine nerve.
g.	Gastrocnemius.	p.m.	Pia mater.
g.d.	Ganglion of nerve-root.	p.n.	Peroneal nerve.
gs.	Gastric branch of vagus.	p.s.c.	Posterior-vertical semicircular canal.
g.t.	Gasserian ganglion.	pt.	Pterygoid muscle.
g.vg.	Ganglion of the trunk of vagus.	pt.b.	Pituitary body.
h.gl.	Harderian gland.	pul.	Pulmonary branch of vagus.
hp.	Hypoglossal (1st spinal) nerve.	r.	Retina.
h.s.c.	Horizontal semicircular canal.	r.b.	Retractor bulbi muscle.
hy.	Hyoid arch.	r.d.	Dorsal ramus of glosso-pharyngeal nerve.
il.	Ilium.	r.d'.	Dorsal ramus of vagus.
i.n.	Internal nostril.	s.	Sacculus.
inf.	Infundibulum.	sc.	Sciatic nerve.
in.r.	Inferior rectus.	scl.	Sclerotic.
i.ob.	Inferior oblique.	s.mb.	Semimembranosus.
ir.	Iris.	s.ob.	Superior oblique.
i.r.	Internal rectus.	sp.n.	Splanchnic nerve.
l.	Crystalline lens.	s.r.	Superior rectus.
lg.	Laryngeal nerve.	st.	Stomach.
l.s.	Lumbo-sacral enlargement.	stp.	Stapes.
l.s.pl.	Lumbo-sacral plexus.	sy.g.	Sympathetic ganglion.
l.tm.	Lamina terminalis.	sy.n.	Sympathetic nerve.
l.v.	Lateral ventricle.	t.	Tongue.
lv.b.	Levator bulbi muscle.	t.a.	Tibialis anticus.
md.	Medulla oblongata.	th.n.	Posterior tibial nerve.
mn.	Mandibular nerve.	th.	Thalamencephalon.
m.ty.	Membrana tympani.	tr.f.	Triceps femoris.
mx.	Maxillary nerve.	ut.	Utriculus.
my.	Myelon.	v.	Vestibule.
na.	Nasal sac.	v'.	Diverticulum of the same.
na'.	Nasal bone.	v.c.	Vitreous chamber.
n.c.	Neural canal.	v³.	Third ventricle.
ol.	"Olfactory lobe."	v⁴.	Fourth ventricle.
ol'.	Ventricle of the same.	vo.	Vomer.
ol.n.	Olfactory nerves.	I. to X.	Cranial nerves I. to X.
o.na.	Orbito-nasal nerve.	1 to 10.	Spinal nerves 1 to 10.
op.	Optic lobe		

PLATE VI.

THE FROG.—GENERAL HISTOLOGY (see Appendix G).

FIG. I.—Ciliated epithelium.
Scraped from the roof of the mouth of a recently killed frog.
Normal salt solution, afterwards stained with eosin. D. 3.

FIG. II.—Columnar epithelium.
Scraped from the lining membrane of the small intestine. Treated as for Fig. I., D. 3.
The four cells to the left retained their natural relationships, and in connection with those numbered i and ii, the mucin secretion is indicated; ii is drawn in the act of discharging the same.

FIG. III.—A portion of the Frog's mesentery, examined in water after two hours' exposure to the sun's rays in silver nitrate solution   D. 2.

FIG. IV.—i. A small piece of intermuscular connective tissue.
Spirit material, stained with borax carmine.
If no white fibres are readily visible, a piece of tendon should be teased up.
ii. Pigment cells (of the mesentery) in various stages of development. D. 4.

FIG. V.—Hyaline cartilage.
i. Thinnest part of the xiphisternum, after removal of its fibrous investment. Fresh, D. 2.
ii. One group of cells of the above, more highly magnified. F. 4.
The nucleus of the uppermost cell has divided, prior to the division of the cell itself.

FIG. VI.—The fat-cell.
i. Section of the corpus adiposum. Alcohol and borax carmine. F. 3.
ii. Cells of the above, teased up fresh. Stained with eosin. D. 3.
iii. A ripe fat-cell. The fat-drops have united to form one large globule.
For an account of the morphology of the corpus adiposum see Marshall (16) and Bourne (3).

FIG. VII.—The blood, examined immediately after death. D. 4.
i. Three phases in the life of the same corpuscle, at successive intervals of two minutes each.
ii. Phases in the union of two corpuscles.
iii. A late stage in the fission of a white corpuscle. F. 3.
At iv. are a number of red corpuscles seen on end.

FIG. VIII.—Transverse section of the spinal cord.
On the right side the pia mater is shown, and more especially the large multipolar nerve-cells of the anterior cornu. A. 3.
Chromic Acid. Alcohol. Borax carmine.

FIG. IX.—Transverse section across the sciatic nerve. Alcohol and hæmatoxylin. A. 3. The sheath, *p.n.*, is often laden with fat.

FIG. X.—A small portion of one of the nerve-bundles of Fig. IX. highly magnified. F. 3.

FIG. XI.—A series of preparations to show the structure of the nerve-fibre, and the leading types of nerve-cells.
  i. A small portion of a nerve-fibre, teased up, and exposed to the action of 1 p.c. osmic acid for two hours. The segment between the two nodes of Ranvier figured, is seen to bear one nerve-corpuscle. D. 2.
  ii. An isolated fibre from a nerve-trunk, examined immediately after death. The lower portion of the preparation is drawn after treatment with 1 p.c. osmic acid. F. 4.
  iii. A small portion of a dead nerve-fibre similar to the above. F. 4.
  iv. A unipolar nerve-cell from the ganglion of a spinal nerve. Teased fresh in eosin. D. 3. These cells vary very much in size; that drawn was a small one.
  v. A bipolar nerve-cell from a sympathetic ganglion. Teased fresh in eosin. F. 2.
  vi. A multipolar nerve-cell from the anterior cornu of the spinal cord (* of Fig. VIII.). This preparation shows the continuation of the nerve-cell into an axis-fibre, and was obtained by teasing up a portion of an anterior nerve-root, isolated together with a small piece of the grey matter of the cord. Fresh, stained eosin. D. 3.
  vii. One of the smaller fusiform nerve-cells from the grey matter of the spinal cord. Teased fresh. Eosin. D. 4.
  In Figs. IV. and V. the sheath is indicated by a faint line.

FIG. XII.—Transverse section of the retina. Picric acid. Alcohol. Borax carmine. F. 3.
The parts shaded darkly are those which stain deepest.
The cells, *n.o.*, usually but two rows deep, may sometimes be three deep as here figured.

FIG. XIII.—Transverse section across the middle of the belly of a small muscle of the hind limb.
For purposes of comparison with the nerve, the section drawn was of the same absolute size as that of the nerve-trunk, Fig. IX. Alcohol and borax carmine. A. 3.
The ingrowths of the perimysium marked * are very delicate, but they may be readily traced as they are generally pigment-laden.

FIG. XIV.—One fibre of the above muscle-section, more highly magnified.
Drawn to the same scale as the nerve-fibres in Fig. X.

FIG. XV.—Two preparations of striped muscular fibre.
  i. A fresh fibre one portion of which has been ruptured, rendering clear the so-called sarcolemma. D. 3.

PLATE VI

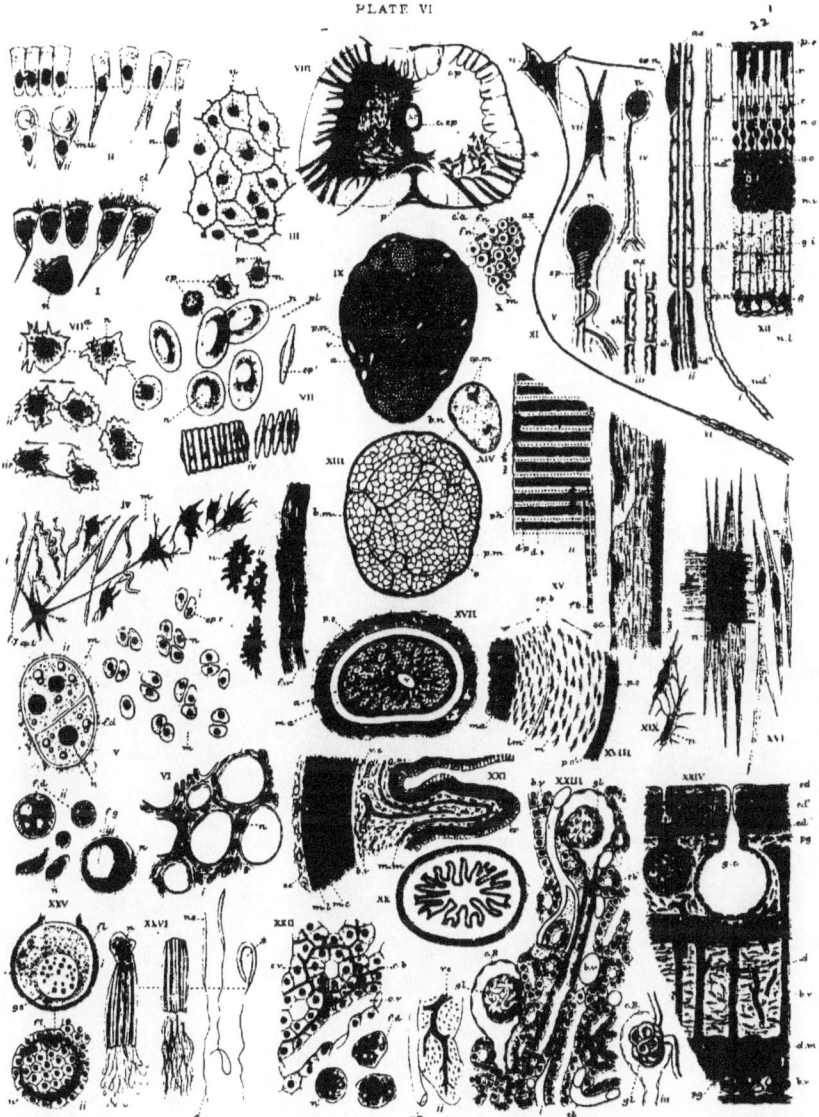

The muscle-corpuscles are drawn as seen after treatment with 1 p.c. acetic acid and magenta; they can, however, be readily observed in the fresh fibre.

ii. A small portion of a similar fibre, examined fresh in normal salt-solution. The so-called fibrils, fb, were obtained by subsequent teasing-up after death.* F. 4.

FIG. XVI.—Unstriped muscular fibre.

Obtained by teasing up a small piece of intestine, after maceration for some days in Müller's fluid. Borax carmine. D. 4.

These cells hang together very tenaciously, and the left-hand portion of the figure represents a small piece of the entire muscular wall of the above-named viscus, the two layers of which overlie each other as in life.

Appearances of this kind often constitute serious sources of error.

FIG. XVII.—Transverse section near the middle of the shaft of the femur.
Decalcified with ½ p.c. chromic acid. Alcohol and borax carmine. A. 2.

FIG. XVIII.—A small portion of the above, more highly magnified.

The middle lamella, lm, separates the periosteal bone from that formed by the marrow; the direction of growth of these is indicated by arrows. D. 3.

FIG. XIX.—Two adjacent bone-corpuscles, from the thinnest portion of the above section, drawn under Gundlach's $\frac{1}{15}$th immersion.

FIG. XX.—An entire transverse-section of the ileum; the muscular and epitheloid layers are shaded darkly.

FIG. XXI.—A small portion of the same, more highly magnified. Alcohol and borax carmine. D. 3.

No note is taken of the mucous drops borne by the epitheloid cells (the goblet-cells of histologists. See Fig. II., i. and ii.)

FIG. XXII.—Section of a small portion of the liver. Alcohol and borax carmine. D. 3.

The bile-capillaries were distended by gently squeezing the biliary fluid back into them from the gall-bladder, the bile-duct being first ligatured.

The isolated cells, drawn under F. 3, were teased out from the fresh liver in salt solution.

---

* The most reliable preparations of striped muscle are to be obtained by the use of the freezing-microtome. The Golding Bird or Pritchard Machines can be recommended. The most recent reliable accounts of the modern aspects of the vexed muscle-striation question are those of Ranvier (19), and Rutherford (34). The system of nomenclature here employed in connection with nerve, muscle, and bone, is Huxley's.

Fig. XXIII.—Three figures illustrating the structure of the kidney.
i. Portion of a section. Alcohol and borax carmine. D. 4.
The blood-vessels indicated are generally full of corpuscles—not here drawn.
ii. The ventral surface of a portion of the right kidney after removal of the peritoneum. Slightly magnified.
The small osculæ represent the apertures of communication (nephrostomes) between the uriniferous tubules and the body-cavity. For details see Spengel (22).
iii.—A glomerulus from a section of the kidney injected from the dorsal aorta. See Appendix D. Alcohol. D. 3.

Fig. XXIV.—Transverse section of skin. Alcohol and borax carmine. D. 3.
Of the two cutaneous glands figured, the left hand one is represented *en face*.

Fig. XXV.—Two young ovicells with their investing follicular epithelium. Alcohol and borax carmine. D. 2.
i. Optical section. ii. Surface view.
The young ovicells such as are here figured, may be easily identified by the absence of that pigment so characteristic of the ripe ovum (see Pl. VI.) They are very small and quite white, and do not always completely fill the follicle as did the one drawn in section above.
For figure of a section of the Frog's ovary see Marshall (16).

Fig. XXVI.—Spermatozon, obtained by teasing up a small portion of the testis.
The head is often carried as at *, thus giving rise to very deceptive appearances. Of the spermatozoan aggregates figured, that of the left side had attached to it a cell of the germinal epithelium. See Bloomfield (1).

a.	Artery.	$c.p.$	Posterior cornu.
$ax.$	Axis fibre.	$cp.b.$	Bone corpuscle.
$b.m.$	Muscle bundle (fasciculus).	$cp.c.$	Cartilage corpuscle.
$b.n.$	Nerve bundle.	$cp.m.$	Muscle corpuscle.
$b.v.$	Blood vessels.	$cp.n.$	Nerve corpuscle.
$c.$	Cones.	$cp.t.$	Connective-tissue corpuscle.
$c.a.$	Anterior cornu.	$c.v.$	Blood capillaries.
$c.b.$	Bile capillaries.	$d.$	Derma.
$c.B.$	Bowman's capsule.	$d.m.$	Muscular layer of derma.
$c.c.$	Canalis centralis.	$d.p.$	Phragmodisc (septal zone).
$c.ep.$	Ciliated epithelium of the same.	$d.s.$	Sarcodisc (interseptal zone).
$cl.$	Cilia.	$ed.$	Superficial horny layer of epiderma.
$cp.$	White corpuscles.	$ed'.$	Middle layer of epiderma.
$cp'.$	Red corpuscles.	$ed''.$	Columnar (Malpighian) layer of epiderma.

ep.	Epithelium.	*n.i.*	Inner nuclear layer.
f.	Flagellum.	*n.l.*	Layer of optic-nerve fibres.
fb.	So-called fibrils.	*n.o.*	Outer nuclear layer.
f.d.	Fat drops.	*n.s.*	Nucleus of spermatozoid.
f.g.	Fat globule.	*p.*	Pia mater.
fl.	Follicular epithelium.	*p.e.*	Pigmented epithelium.
f.n.	Medullated nerve fibre.	*p.g.*	Pigment cells.
f.n'.	So-called non-medullated nerve fibre.	*ph.*	Sarcophragm (septal line).
f.w.	White connective-tissue fibres.	*pl.*	Plasma.
f.y.	Yellow connective-tissue fibres.	*p.m.*	Perimysium.
g.	Layer of ganglion cells.	*p.n.*	Perineurium.
g.c.	Cutaneous glands.	*po.*	Periosteum.
g.i.	Inner granular layer.	*p.o'.*	Osteogenic layer of the same.
gl.	Glomerulus.	*ps.*	Pseudopodium.
g.o.	Outer granular layer.	*r.*	Rods.
g.s.	Germinal spots.	*r.e.*	Efferent renal vessels.
g.v.	Germinal vesicle.	*sc.*	Sarcolemma.
lm.	Middle lamella.	*se.*	Serous (peritoneal) layer.
m.	**Matrix.**	*sh.*	Primitive sheath.
ma.	Marrow cells.	*sh'.*	Medullary sheath.
m.c.	Circular muscular layer.	*sp.*	Processus spiralis.
m.l.	Longitudinal muscular layer.	*th.*	Uriniferous tubule.
m.m.	Muscular layer of mucous membrane.	*tb'.*	Ciliated segment of the above.
mu.	Mucin drop.	*ur.*	Ureter.
n.	Nucleus.	*v.*	Vein.
nd'.	Ranvier's node.	*v.c.*	Vascular connective tissue.
nd".	Schmidt's node.		

PLATE VII

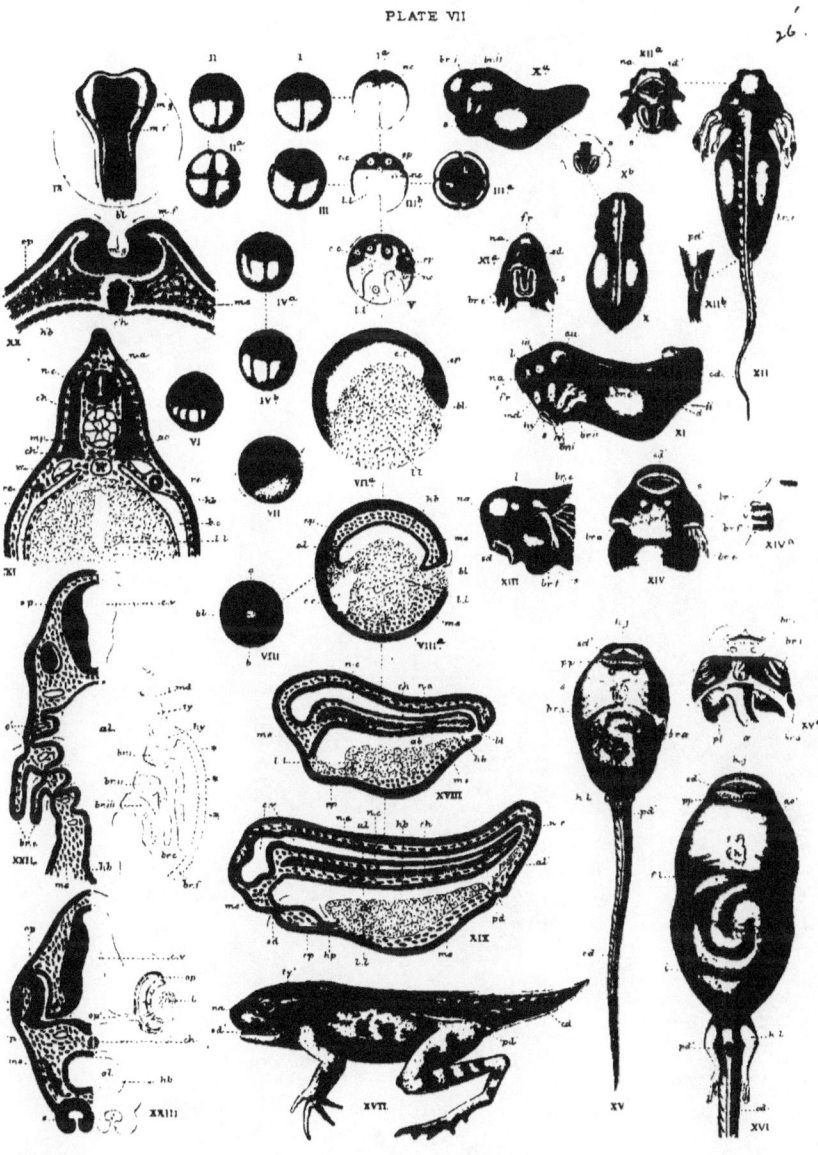

PLATE VII.

THE FROG.—Embryology and Larval Metamorphoses.

Fig. I.—The segmenting ovum, on the appearance of the first cleavage furrow.

Fig. I*.*—Section of the same, at right angles to the furrow.

Fig. II.—The same, on the appearance of the second furrow, viewed slightly from above.

Fig. II*.—The above, seen from beneath.

Fig. III.—The same, on the appearance of the third—first horizontal—furrow.

Fig. III*.—The same, seen from above.

Fig. III*b*.—Longitudinal section of the same.

Figs. IV*a*. and IV*b*.—Two phases in the segmentation of the ovum, on the appearance of the fourth and fifth furrows.

Fig. V.*—Longitudinal vertical section, at a slightly later stage than the above.

Fig. VI.—A later stage. The upper pigmented pole is seen to be dividing more rapidly than the lower.

Fig. VII.—A still later phase in the same.

Fig. VII*.—Longitudinal vertical section of Fig. VII.

Fig. VIII.—The segmenting ovum, at the blastopore stage.

Fig. VIII*.—Longitudinal vertical section of the same.

(Figs. VII*. and VIII*. × 10. All the others × 5.)

Rauber (20) has recently made an elaborate study of the details of the early segmentation of the Frog's ovum.

Fig. IX.—Surface view at a stage somewhat later than Fig. VIII.
The medullary folds are seen to be approximating at one point; they finally meet there and continue to close in opposite directions, as indicated by the arrows. × 10.

Fig. X.—The young tadpole, at the first appearance of visceral arches. Dorsal view.

* The nuclei of these figures are diagrammatic; they are only occasionally visible in cells so large.

FIG. X^a.—The same, from the left side.

FIG. X^b.—The anterior end of the above, ventral view.

FIG. XI. The tadpole, at the first appearance of external gills, body segments and the organs of the higher senses.
The animal is at this period still enclosed within the egg-membranes, the tail being curved to one side as indicated.

FIG. XI^a.—The head of the same, ventral view.

FIG. XII.—The tadpole, on the assumption of the free swimming stage. The external gills are at their maximum. Dorsal view.

FIG. XII^a.—The head of the same seen from beneath.

FIG. XII^b.—The root of the tail of the same, ventral view.

FIG. XIII.—The head of a tadpole, on the appearance of the opercular fold, seen from the left side.

FIG. XIV.—The tadpole's head, during the period at which external and internal gills coexist. Ventral view.

FIG. XIV^a.—The same, the branchial chamber being opened up on the right side.

FIG. XV.—The tadpole, at the period in which the last remnants of suckers and first traces of hind limbs are visible. Ventral view.

FIG. XV^a.—The same.
Both branchial chamber and body-cavity having been opened up, showing that internal gills and lungs coexist.

FIG. XVI.—A late stage of the frog's tadpole.
Both fore and hind limbs are visible, the former still buried up beneath the larval integument. Ventral view.

FIG. XVII.—The late larva, on the assumption of a prehensile-mouthed exclusively air-breathing stage. The tail was partly absorbed. Seen from the left side.
(Figs. X. to XVII. all × 5.)

FIG. XVIII.—Longitudinal vertical section of the embryo, at a stage slightly earlier than that of Fig. X. × 10.

FIG. XIX.—A similar section of a later tadpole.*
From nature, after Goette (8). × 10.

* The blastopore is stated by Spencer (Zool. Anz., February 23rd, 1895) to persist as the anus.

Fig. XX.—Transverse section of Fig. IX., to show the mode of origin of the nervous axis. D. 2.

Fig. XXI.—Transverse section, taken across the middle of Fig. XI.

Owing to the tendency of the body-wall to shrink under the reaction of reagents, the body-cavity, *b.c.*, rarely appears as spacious as it is here figured.

On the left side a slightly earlier stage in the development of the prorenal duct, *re*, is represented, than would be seen at this period. See Fürbringer (7), and Sedgwick (21). D. 2.

Fig. XXII.—The anterior portion of a section, through the plane i.—ii. of Fig. XI., showing the origin of the visceral clefts and arches, and of the external gills.

The right half of the figure indicates, in outline, the same parts at a later stage; those clefts which open up are thus represented, and the opercular fold and a third external gill which had by this period come into existence, are both indicated by dotted lines. D. 2.

The surfaces marked thus * are those which eventually bear the internal gills.

Fig. XXIII.—A section across the plane iii.—iv. of Fig. XI., to show the mode of origin of the essential parts of the eye, taken as a type of the higher sense organs.

The right half of the figure illustrates the same at a later stage, after the lens has lost all connection with the epiderma. D. 2.

Note.—Only the initial stages in the segmentation of the ovum are here figured, and the central row of drawings, connected together by a longitudinal line, represents sections through the same plane of a progressive series.

Except in Fig. IX., the arrows follow the direction of growth of the cells at the point indicated.

al.	Archenteron (midgut).		*cd.*	Caudal appendage (tail).
al'.	Postanal gut.		*ch.*	Notochord.
ao.	Aorta.		*ch'.*	Subnotochordal rod.
ao'.	Aortic arches.		*c.v.*	Cerebral vesicle (brain).
au.	Auditory involution.		*ep.*	Epiblast.
b.c.	Body cavity.		*f.l.*	Fore limb.
bl.	Blastopore.		*fr.*	Fronto-nasal process.
br.	Branchial arches.		*h.*	Heart.
br.a.	Branchial aperture.		*h.j.*	Horny jaw.
br.c.	Branchial chamber.		*h.l*	Hind limb.
br.e.	External gills.		*hp.*	Hepatic diverticulum.
br.f.	Branchial fold (operculum).		*hy.*	Hyoid arch.
br.i.	Internal gills.		*i.*	Coiled intestine.
c.c.	Cleavage cavity.		*l.*	Crystalline lens.

l.l.	Yolk-bearing lower-layer cells.	pd'.	Aperture of involution of the same (cloacal aperture).
md.	Mandibular arch.		
m.f.	Medullary fold.	pl.	Pulmonary sac (lung).
m.g.	Medullary groove.	pp.	Oral papillæ.
w.p.	Muscle plate.	re.	Prorenal (segmental) duct.
ms.	Undifferentiated mesoblast.	s.	Head suckers.
na.	Nasal involution (anterior nostril).	sd.	Stomodæum.
n.a.	Neural (cerebro-spinal) axis.	sd'.	Aperture of involution of the same (oral aperture).
nc.	Nucleus.		
n.c.	Neural canal.	sm.	Somatic mesoblast.
n.e.	Neurenteric canal.	sp.	Splanchnic mesoblast.
œ.	Œsophagus.	ty.	Tympano—Eustachian cleft.
op.	Optic cup (retina).	ty'.	Membrana tympani.
op'.	Optic stalk (optic nerve).	v.c.	Posterior cardinal vein.
pd.	Proctodæum.		

THE CRAYFISH.

PLATES VIII. to X.

THE CRAYFISH.

PLATES VIII. TO X.

THE CRAYFISH.

a.	Anus.		c.l.	Longitudinal nerve-commissure.
a'	Anal valve.		c.m.	Muscles of chela (basal joint of).
a.a.	Antennary artery.		cn.	Cornua.
ab.	Abductor muscle.		cœ.	Cœcum.
ab. 1-6.	Abdominal appendages, 1-6.		c.ov.	Ovoid corpuscle.
a.b.	Arthrobranchia, external set.		cp.	Carpopodite.
a.b'.	Arthrobranchia, internal set.		c.p.	Cardio-pyloric muscle.
abd.	Abdomen.		c.p'.	Posterior transverse nerve-commissure.
a.c.	Alary muscles.		c.r.	Crystalline rods.
ad.	Adductor muscle.		ct'.	Cuticular lining of fore gut.
ad.m.	Adductor muscle of mandible.		ct".	Cuticular lining of hind gut.
a.g.	Anterior gastric muscle.		cv.g.	Cervical groove.
a.h.	Auditory hair.		c.c.	Coxopodite.
al.	Archenteron.		c.r.s.	Coxopoditic setæ.
a.m.	Articular membrane.		dc.	Dactylopodite.
amb.	Ambulatory legs.		d.g.	Depressor muscles of stomach.
an.	Antennary muscles.		dl.g.	Dilator muscles of stomach.
an'.	Antennule.		e.	Epistoma.
an".	Antenna.		eb.	Epiblast.
a.p.	Abdominal papilla.		ec.o.	Ectostracum.
ar.	Articular facet.		en.	Endopodite.
au.	Auditory sac.		en.o.	Endostracum.
au'.	External aperture of the same.		e.o.	Epiostracum.
bc.	Branchiostegite.		ep.	Epipodite.
b.c.	Branchio-cardiac trunks.		e.pl.	Endopleurite.
b.ch.	Branchial chamber.		epm.	Epimeron.
b.d.	"Bile-duct."		e.s.	Eyestalk.
bl.	Blastopore.		e.st.	Endosternite.
br.	Gill.		ex.	Exopodite.
br.a.	Afferent branchial vessel.		ex.a.	Extensor abdominis muscle.
br.c.	Branchio-cardiac groove.		f.e.	Follicular epithelium of ovisac.
br.e.	Efferent branchial vessel.		f.g.	Fore gut.
bs.	Basipodite.		fl.a.	Flexor abdominis muscle.
c.	Carapace.		f.t.	Fatty connective tissue.
c.a.	Anterior transverse nerve-commissure.		g.	Gastrolith.
c.an.	Annulate corpuscle.		g.a.	Genital aperture.
cd.	Cardiac chamber.		g.ab.	Abdominal ganglia.
cd'.	Dorsal cardiac ossicle.		g.c.	Cerebral ganglion.
cd".	Median cardiac ossicle.		g.g.	Green gland, aperture of.
cd.l.	Lateral cardiac ossicle.		g.œ.	Subœsophageal ganglion.
c.g.	Constrictor muscles of stomach.		gn 1-20.	Ganglia 1-20.
ch.	Chela.		g.op.	Optic ganglion.

5

g.s.	Germinal spots.	ot.	Otoliths.
g.v.	Germinal vesicle.	ov'.	Ovary, anterior lobe.
h.	Heart.	ov".	Ovary, posterior lobe.
h.d.	Hypodermis.	p.	Pericardial sinus.
h.g.	Hind gut.	p.b.	Pleurobranchia.
hp.	Hepatic artery.	p.b'.	Rudimentary pleurobranchia.
h.pe.	Digestive gland (so-called liver).	pc.	Ferment cells.
hy.	Hypoblast.	pc'.	Secretionary product of the same.
i.	Intestine.	p.c.	Procephalic process.
i.a.	Inferior abdominal artery.	p.ca.	Pore canals.
is.	Ischiopodite.	p.g.	Posterior gastric muscles.
i.s.	Intersternal membrane.	pl.	Pleuron.
i.t.	Intertergal membrane.	pr.	Propodite.
lb.	Labrum.	pt.	Protopodite.
l.b.	Lenticular bodies.	py.	Pyloric chamber.
l.c.	Liver cells.	py'.	Dorsal pyloric ossicle.
l.g.	Levator muscles of stomach.	py".	Median pyloric ossicle.
l.t'.	Lateral tooth.	py.l.	Lateral pyloric ossicle.
l.t."	Accessory lateral tooth.	r.	Rectum.
l.r.	Levator abdominis muscle.	re.	Green gland.
m.	Mouth.	re'.	Vestibule of the same.
mh.	Mesoblast cells.	s.	Stomach.
md.	Mandible.	s.a.	Superior abdominal artery.
me.	Meropodite.	se.	Setæ.
m.g.	Midgut.	sy.	Scaphognathite.
m.m.	Mandibular muscles.	s.s.	Sternal sinus.
m.op.	Muscles of eye-stalk.	s.sp.	Striated spindle.
m.p.	Mandibular palp.	st.	Sternal artery.
m.pp.	Membrana propria.	st 1-20.	Sterna 1 to 20.
mt.	Metastoma.	sw.	Swimmerets.
m.t.	Median tooth.	t.	Telson.
m.r 1-2.	Maxillæ 1 and 2.	t.f.	Tail fin.
m.rp 1-3	Maxillipedes 1 to 3.	tg 1-20.	Terga 1 to 20.
n.au.	Auditory nerve.	th.a.	Thoraco-abdominal linkwork.
ne.	Nucleus.	ts'.	Testis, anterior lobe.
n.g.	Ganglionic nerves.	ts".	Testis, posterior lobe.
n.ig.	Interganglionic nerves.	v'.	Dorsal valves of heart.
n.op.	Optic nerve.	v".	Ventral valves of heart.
n.v'.	Anterior visceral nerve.	v"'.	Lateral valves of heart.
n.v".	Posterior visceral nerve.	v.cp'.	Lateral cardio-pyloric valve.
o.c.	Ovicell.	v.cp".	Median cardio-pyloric valve.
o.d.	Ovary duct.	v.d.	Vas deferens.
o.d'.	Cut edge of the same.	v.p.	Valvular termination of lining of fore-gut.
œ.	Œsophagus.	y.	Food yolk.
ol.	"Olfactory" setæ.	i. to xx.	Appendages i. to xx.
op.	Ophthalmic artery.	i'. to xx'.	Articulations for the same.
o.s'.	Ovisac, ripe.	1. to 20.	Segments 1 to 20.
o.s".	Ovisac, young.		

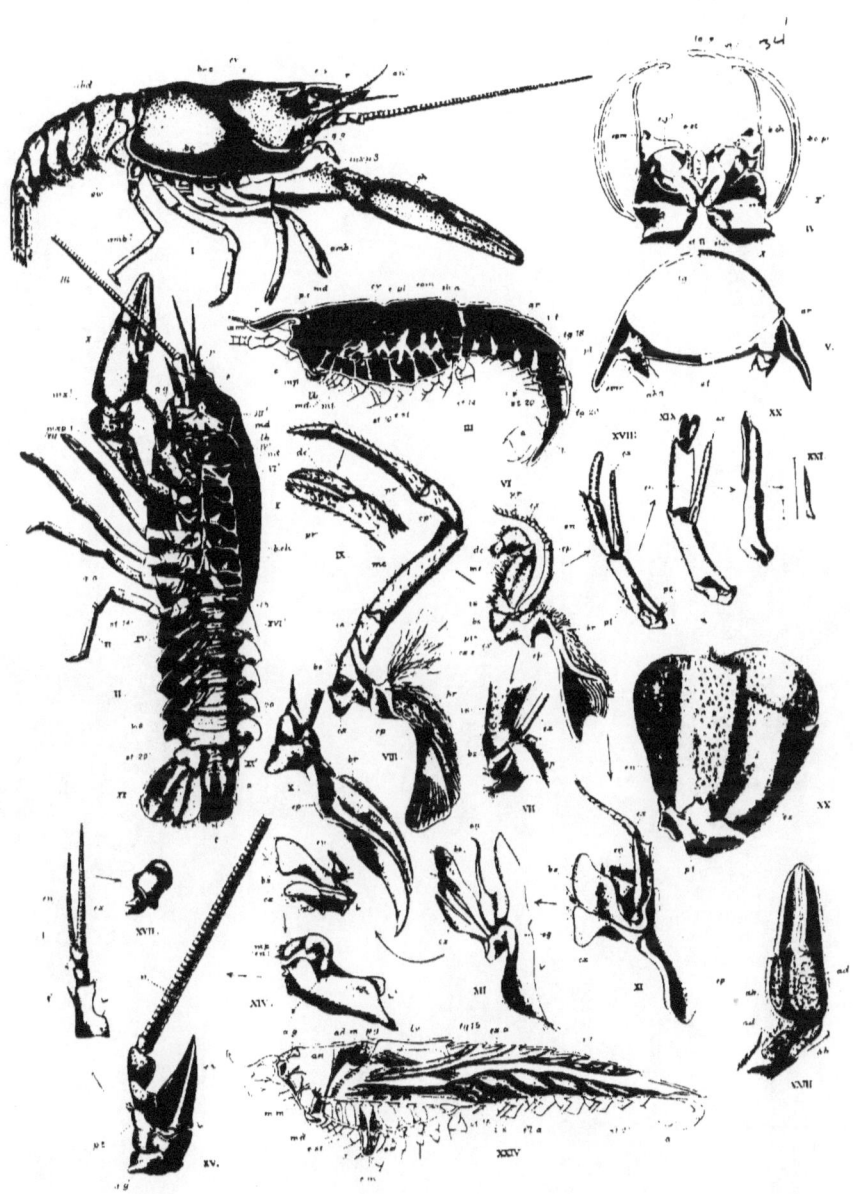

PLATE VIII.

THE CRAYFISH.†—THE EXOSKELETON AND APPENDAGES. THE MUSCULAR SYSTEM.

FIG. I.—The entire animal, from the left side, in the resting attitude ♀. Nat. size.

FIG. II.—The same from beneath, all the appendages of the left side and a portion of the branchiostegite having been removed, in order to display the whole sternal surface.

The last ambulatory leg is drawn in its functional position, as the chief lever in forward progression. The feeler of the antenna is indicated as underlying the chela, in error. Nat. size.

FIG. III.—Median vertical-longitudinal section of the same, the soft parts having been removed, after boiling in 10 p.c. potash solution. Nat. size.

FIG. IV.—Transverse section of the thorax, taken immediately behind the bases of the great chelæ, treated as in Fig. III.

The small ingrowths marked thus * give part attachment to the levator muscles of the fore-gut, and may be appropriately termed *endotergites*. × 2.

FIG. V.—The third abdominal segment of the ♀, seen from behind.
On the left side the parts are represented in section, for purposes of comparison with Fig. IV. × 2.

FIGS. VI. to XXII.—The appendages. All × 3.
Only the typical ones are drawn. Taking the second maxillipede (Fig. VI.) as a central type, the others are represented in series, the divergent lines of which are indicated by arrows.

All are drawn in the relatively natural position, their posterior surfaces being alone represented.

VII.—A portion of the third maxillipede, to show the fusion of the ischiopodite and basipodite, characteristic of it and the great chela.

VIII.—The penultimate ambulatory appendage.

IX.—The two terminal joints of the first ambulatory leg.

X.—The gill-bearing portion of the Lobster's third ambulatory appendage.

XI.—The first maxillipede of the Crayfish.

* All the figures of Plates VIII. to X., unless otherwise stated, refer to the Red-footed French Crayfish (*Astacus fluviatilis*, var. *nobilis*), specimens of which can often be obtained in our English markets 5 to 6 inches in length.

XII.—The second maxilla.

XIII.—The first maxilla.
The coxopodite of this appendage is carried, during life, inserted into the mouth. (Compare Plate IX., Fig. V.)

XIV.—The mandible.
The endopodite of the second maxilla, Fig. XII., *en*, is carried, during life, hooked over the recess marked*.

XV.—The antenna.

XVI.—The antennule.

XVII.—The eye-stalk.

XVIII.—The third abdominal appendage of the ♀, taken as a type of that series.

XIX.—The second abdominal appendage of the ♂.

XX.—The first abdominal appendage of the ♂.

XXI.—The same of the ♀.
This limb varies very much in size in different individuals; the extremes observed are indicated by the two lines drawn to the left.

XXII.—The last abdominal appendage.

FIG. XXIII.—The three terminal joints of the right chela, laid open, to display their muscles.
The dotted lines indicate the positions of the underlying tendons. Nat. size.

FIG. XXIV.—Longitudinal section of the whole animal, all the soft parts having been removed, with the exception of the muscles.
Portions of the great gastric muscles (compare Fig. VII., Plate IX.) and of those connected with the basal joints of the chela, antenna, and mandible, have been left. Nat. size.

This dissection is best performed upon an animal previously hardened in alcohol.

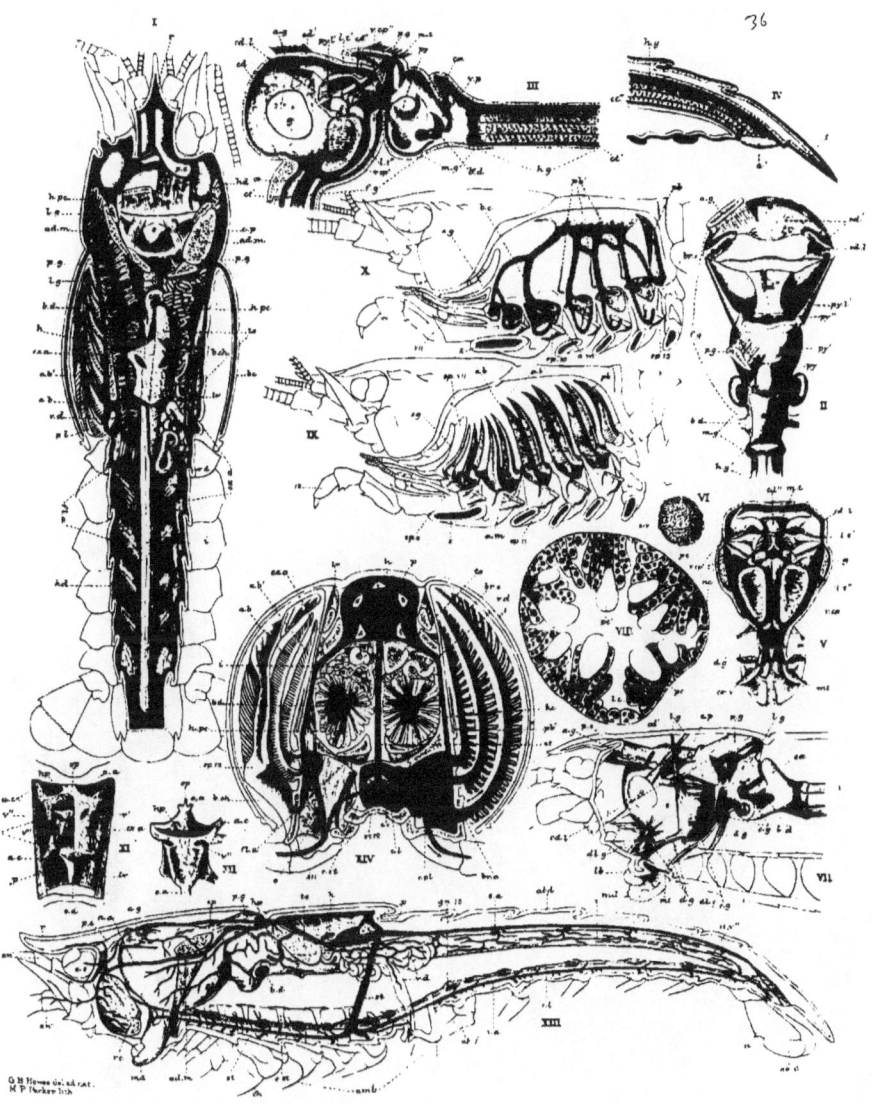

PLATE IX.

THE CRAYFISH.—THE GENERAL DISPOSITION OF THE VISCERA. THE ORGANS OF ALIMENTATION, CIRCULATION, AND RESPIRATION.

FIG. I.—A general dissection from the dorsal aspect.

The whole tergal skeleton has been removed. The extensor-abdominis muscle, the greater part of the generative apparatus, and part of the anterior gastric muscle together with the procephalic process, have been removed on the left side; the adductor muscle of the mandible has been cut short, and the digestive gland (so-called liver) dissected away to the level of its central duct.

On the right side the gills were removed, to show the extent of the branchial chamber. δ. $\times 1\frac{1}{2}$.

FIG. II.—The stomach and anterior portion of the intestine, after removal from the body, seen from above.

The attachments of the gastric muscles (approximators of the gastric teeth) are shown on the left side.

The whole fore-gut was divested of its outer cellular coat.

FIG. III.—The same, in median longitudinal section, the wall of the fore-gut being left entire.

FIG. IV.—The terminal portions of the hind-gut and abdomen, in longitudinal section.

FIG. V.—Transverse section of the fore-gut, passing through the œsophagus immediately behind the anterior gastric muscle; a young gastrolith, $g.$, was present.

The metastoma and basal joint of the first maxilla are drawn in natural position.

$cx.v.$ should read $bs.v.$ in this figure.

In Figs. III. to V. the cut edges of the cuticular lining and its related structures are drawn in black.

FIG. VI.—Outer view of the gastrolith.

(Figs. II. to VI., all $\times 2$.)

FIG. VII.—Dissection from the left side, to show the leading muscles of the fore-gut, and—incidentally in section—the labrum, a portion of the metastoma and the procephalic process.

The specimen was previously hardened in alcohol. $\times 1\frac{1}{2}$.

In dissecting the above, the opportunity should not be lost of noting the extent to which the calcifications of the chitinous wall of the stomach are determined by muscular attachment.

Fig. VIII.—Transverse section of one of the digestive cœca.

The preparation drawn was frozen in gum, by means of the "Golding Bird" ether microtome; the sections were transferred as cut to 1 p.c. osmic acid solution, the fatty globules of the liver-cells (*l.c.*) being thus stained black, as drawn. D. 3.

Weber (37), who first fully elucidated the real nature of this so-called liver, significantly termed it the " hepato-pancreas."

Fig. IX.—Dissection from the left side, to show the relations of the gills.

The branchiostegite was removed, the larger appendages were cut short, and only sufficient of the bases of the branchiæ borne by them were left to indicate their positions. (Typical podobranchiæ are figured, in detail, on Plate VIII.)

The shaded areas represent the interarticular membranes, *a.m.* × 1½.

Fig. X.—The same, after injection of the branchio-cardiac trunks, and their factors— the efferent branchial vessels.

All the branchiæ were removed, excepting the solitary functional and the three vistigial pleurobranchiæ, *p.b.* These were, in this specimen, at their maximum of development. × 1¼.

In injecting, the branchiostegite was first dissected off, and a small hole drilled in the carapace from above; through this the syringe was inserted into the pericardial sinus, injection being completed before the removal of any of the gills.

Fig. XI.—The pericardial sinus, the heart and related structures, as seen from above.

The pericardium was opened up, after 24 hours in alcohol; the left side of the heart was removed to the level of its lateral valve, and sufficient of the extensor abdominis muscle dissected out to render clear the osculæ of the branchio-cardiac trunks.

Fig. XII.—The heart, after removal from the pericardium, seen from the ventral aspect.

The hepatic artery of the left side was torn away at its base; only its aperture therefore is indicated.

(Figs. XI. and XII. × 1½.)

Fig. XIII.—The arterial system and pericardial sinus, seen from the left side.

The specimen was pinned down in the position drawn, sufficient of the carapace being removed to expose the lateral cardio-pericardiac valve; from the aperture bounded by this injection was performed.

With the exception of the last ambulatory limb, the genital duct related thereto and the posterior gastric muscle, the entire postoral portion of the body was, on the left side, removed to the level of the middle line and with it the digestive gland. All the organs in front of and including the mandible, were left untouched.

The pericardial sinus—the limits of which are indicated in deep black—was carefully dissected after two days' immersion in alcohol. *s*. × 1½.

All the arteries give off—in addition to those branches here figured—small ones to immediately adjacent parts.

FIG. XIV.—Transverse section across the thorax, between the twelfth and thirteenth appendages, after injection as described for Fig. XIII.

Specially designed to show the great blood passages in relation to the gills.

On the left side all the parts are drawn in situ, the muscles and gills included.

On the right the muscles connected with the basal joint of the second ambulatory leg were removed, to fully expose the sternal sinus.

The gills of the right side are diagrammatic, and no note has been taken of the blood-spaces surrounding the extensor and levator-abdominis muscles.

The arrows indicate the course taken by the blood in life.

The central duct of the digestive gland is best seen after preservation in alcohol. *d*. × 3.

The sternal artery is represented in Figs. XIII. and XIV. as passing to opposite sides of the intestine. There is no rule in this matter.

To complete the study of the blood-vascular system, the circulation should be witnessed under the microscope in some small transparent form. *Asellus* answers admirably.

PLATE X.

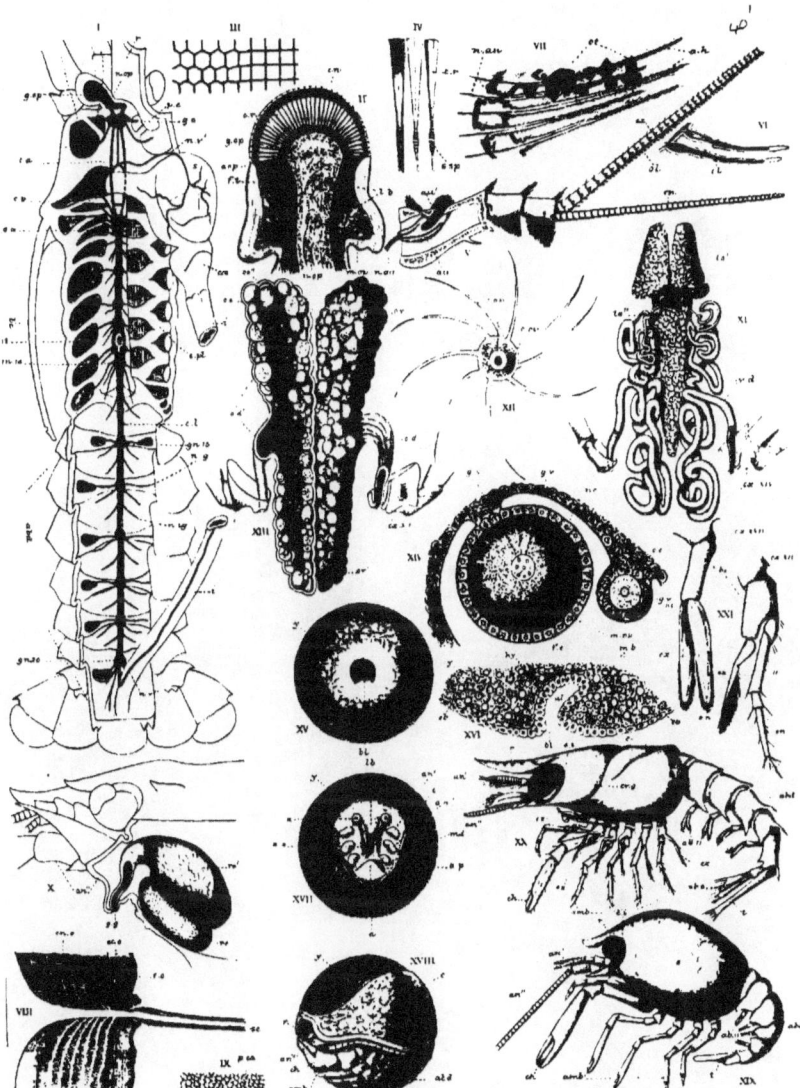

PLATE X.

THE CRAYFISH.—NERVOUS SYSTEM AND SENSE ORGANS. THE HISTOLOGY OF THE EXOSKELETON. THE RENAL AND REPRODUCTIVE ORGANS, AND DEVELOPMENT.

FIG. I.—The nervous system, dissected from above.

The dorso-lateral skeleton, together with the large muscles and the viscera—exclusive of the alimentary canal—were removed; the intestine was severed in order that the visceral nerves might be seen.

On the right side the endophragmal system was left untouched, while on the left it and all else were removed, to the level of the articulations of the appendages; these are indicated by a shade. The left half of the rostrum was dissected away, and the eye-stalk opened up. *s.* × 2.

This figure illustrates the relations borne by the nervous system to the body segments; having made that out, the student is recommended to remove the whole nerve-chain and examine it under water on a black surface.

FIG. II.—Median longitudinal section, through the eye stalk. Chromic acid preparation. A. 2.

The zones of the rod and spindle layer shaded over, are those at which pigment is most fully developed.

FIG. III.—A portion of the cornea, stripped off and examined under water. D. 3.

This figure is an accurate drawing of a portion in which the ordinary typical square facets graduate into hexagonal ones, both types being present in *Astacus*.

FIG. IV.—Three of the crystalline bodies from a fresh eye, teased up in water.

The middle one is drawn *en face*, the right hand one from the side; the pigment is indicated in connection with the left hand one alone. D. 3.

FIG. V.—The left antennule, dissected *in situ* from within.

The inner wall was removed to the level of the auditory sac, which was in part opened up. × 6.

FIG. VI.—Two of the exopoditic sensory hairs of the above, drawn in the natural position, the exopodite being in life directed upwards as in Fig. V. (Compare Fig. I., Plate VIII.) D. 3.

FIG. VII.—A small portion of the auditory sac, viewed from within, after removal from the body.

The sandy particles functional as otoliths, are indicated in the upper portion of the figure alone. The line of attachment of the auditory hairs coincides with that described by the nerve, *n.au.*, of Fig. V. D. 3.

Fig. VIII.—A small portion of a thin transverse section across the ischiopodite of the third maxillipede.
The pore-canals are drawn in the upper half of the figure alone; they are closer together than here represented. D. 3.

Fig. IX.—Portion of a tangential section of the same.
These sections were ground down on a rough surface. See Appendix H.
The appendage chosen for the above is an instructive one, as all gradations of the exoskeleton, from the seta to the cutting tooth, can readily be obtained in the same slice.

Fig. X.—The excretory organ of the left side, dissected from without, a portion of its duct of communication with the exterior having been opened up. × 2.
Wassiliew (36) states that the green gland consists of an unbroken tube of three segments.

Fig. XI.—The male reproductive organs, drawn from above.
The right limb was straightened out to show the whole course of the vas deferens, a portion of which was opened up. × 3.

Fig. XII.—A ripe spermatozoon. D. 3. Fresh.

Fig. XIII.—The female reproductive organs, dissected as for Fig. XI.
The central cavity of the ovary has been laid bare on the left side. × 2.
Figs. XI. and XIII. were drawn from sexually mature specimens, and therefore represent the organs as they appear during the breeding season.

Fig. XIV.—Two stages in the development of the ovisac. D. 3.
A small portion of the germinal epithelium was selected, as described for that of the Frog (Plate VI., Fig. XXV.), and placed on a slide in magenta solution. To avoid undue pressure, a cover-slip was placed on either side of the preparation, upon which to support that covering the object. Modified from Huxley (29).

Fig. XV.—Surface view of the Crayfish ovum at the close of segmentation (a stage equal to that of the Frog, Fig. VIII., Plate VII.) 2 inch objective.

Fig. XVI.—Longitudinal section of a portion of the same, through a plane corresponding with that of the reference line, *bl*, of Fig. XV. D. 2.
For further details see Reichenbach (33).

Fig. XVII.—A surface view of the embryo at an early stage (the equivalent of the free-swimming "Nauplius" stage of the typical Crustacean development).

Fig. XVIII.—Side view of an older embryo. After Rathke (32).

Fig. XIX.—The advanced Crayfish embryo, just prior to the first moult; drawn after removal from the swimmerets of its mother.

The successive stages of the Crayfish development are figured and fully described in Rathke (32) and Huxley (29).

Fig. XX.—An embryo Lobster, shortly after the appearance of its swimmerets.

Sars (35) has published excellent drawings of all the larval metamorphoses of this animal.

(Figs. XVII. to XX. were drawn under a two-inch objective.)

Fig. XXI.—The third abdominal and sixth thoracic appendages (the latter lettered ii.) of the above, drawn to the same scale. A. 2.

THE EARTHWORM.

PLATES XI., XII.

PLATES XI., XII.

THE EARTHWORM.

a.	Anus.	m.l.	Longitudinal muscles of body-wall.
bc.	Buccal sac.	m.l'.	Longitudinal muscles of alimentary canal.
bc'.	Muscles of the same.	m.r.	Radiating fibres of mesentery.
bl.	Passive portion of spermatozoid forming body.	m.s.	Mesenteric septum.
		n.a.	Circumneural arcade.
c.	Cuticle.	nc.	Nucleus.
cy.	Clitellum.	nc'.	Corpuscles of red blood.
c.l.	Longitudinal nerve-commissure.	nc".	Heads (nuclei) of developing spermatozoa.
cm.	Circular blood vascular commissure.	n.c.	Nerve-cells.
cm'.	Interneural blood vascular commissure.	n.g.	Ganglionic nerves.
c.œ.	Circumœsophageal nerve-commissure.	n.ig.	Interganglionic nerves.
e.a.	Epitheloid lining of alimentary canal.	n.l.	Lateral neural vessel.
e.a'.	Subepitheloid vascular layer of the same.	n.s.	Supraneural vessel.
e.c.	Ciliated epithelium.	n.s'.	Subneural vessel.
e.m.	Mesentery of excretory organ.	od.	Oviduct.
ep.	"Hypodermis."	od'.	Internal aperture of the same.
ep'.	Cells of the same.	od".	External aperture of the same.
es.	Epithelium of excretory (segmental) organ.	œ.	Œsophagus.
e.v.	Blood-vessels of excretory organ.	œ'.	Crop.
e.v'.	Excretory plexus.	œ.g.	Œsophageal (calciferous) glands.
fl.	Flagellum.	œ.g'.	Apertures of the same.
g.œ.	Supraœsophageal ganglion.	œ.l.	Lateral œsophageal blood-vessel.
g.i.	First ventral ganglion.	ov.	Ovary.
gl.	Cutaneous glands.	ov'.	Young ova.
gl'.	So-called capsulogeneous glands.	or".	Ripe ovum.
gl".	Gland at base of vas deferens.	p.	Peritoneal membrane.
g.n.	Ganglionic swellings.	p.d.	Dorsal pore.
g.v.	Germinal vesicle.	pg.	Pigment.
gz.	Gizzard.	ph.	Pharynx.
h.	So-called "hearts."	ph'.	Muscles of the same.
hp.	So-called "liver."	ph".	Sucker-like fold of pharynx.
i.	Intestine.	p.m.	Protractor muscles of setæ.
i'.	Intestinal sacculæ.	p.s.	Prostomium.
i.l.	Lateral intestinal blood-vessel.	pr.s.	Peristomium.
i.l'.	Loop of hepatic tissue surrounding the same.	r.m.	Retractor muscles of setæ.
		s.	Functional setæ.
i.s.	Supra intestinal vessel.	s".	Follicle, bearing young setæ.
i.s'.	Subintestinal vessel.	sg.	Segmental organ.
m.	Mouth.	sg'.	Its thin-walled segment.
m.c.	Circular muscles of body wall.	sg".	Its thick-walled segment.
m.c'.	Circular muscles of alimentary canal.	sg'".	Its muscular segment
m.c".	Circular fibres of mesentery.	sg.e.	External aperture of segmental organ.

ATLAS OF BIOLOGY.

sg.i.	Internal aperture of segmental organ.	*tp.*	Typhlosole.
sh.	Integumentary sheath of seta.	*ts.*	Testis.
sh'.	Cuticular sheath of seta.	*v.d.*	Vas deferens.
s.l.	Lateral setæ.	*v.d'.*	Internal aperture (mouth or ciliated rosette) of the same.
sm.	Somitic constriction.		
s.n.	Muscular sheath of nervous system.	*v.d".*	External aperture of the same.
s.n'.	Tubular fibres of the same.	*v.d'".*	Coiled loop of the same.
sp.	Spermathecæ.	*v.s.*	Vesiculæ seminales.
sp'.	Apertures of the same.	*zn.*	Zonitic constrictions.
s.s.	Sac of setæ.	1. to 86.	Body segments 1 to 86.
s.v.	Ventral setæ.		

PLATE XI

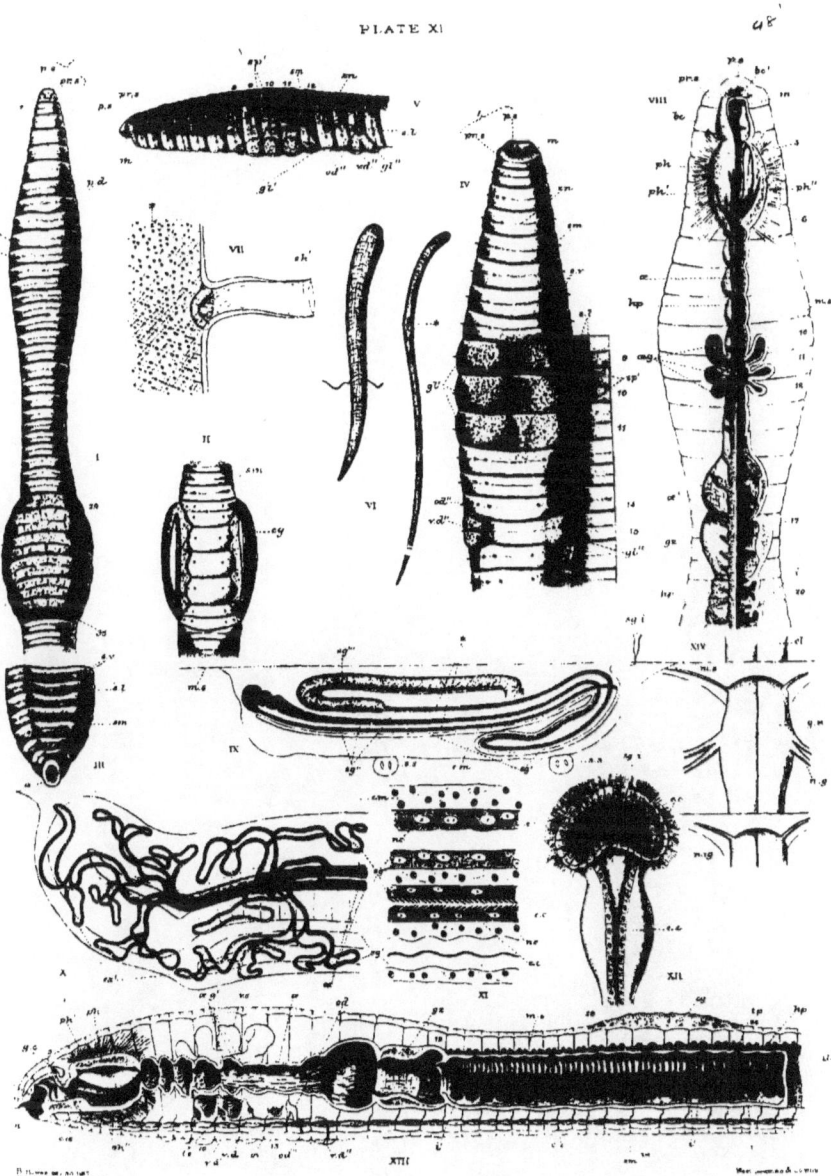

PLATE XI.

THE EARTHWORM.—THE EXTERNAL CHARACTERS. THE ALIMENTARY CANAL AND ITS APPENDAGES. THE EXCRETORY ORGAN, ETC.

FIG. I.—Dorsal view of the anterior end of the body of a sexually mature worm.

The zonitic markings, *zn.*, may often be more than one in number for each segment, especially in the generative region.*

FIG. II.—Ventral view of a portion of the same.

FIG. III.—Ventral view of the terminal segments of the same, the anus slightly upturned.

Figs. I. to III., all × 2.

FIG. IV.—Ventral view of the first seventeen segments, after removal of the cuticle; a portion of the integument of the left side has been pinned out, to show the openings of the spermathecæ.

The apertures of the segmental organs are not indicated. (Compare Fig. V., Plate XII.) × 3.

FIG. V.—The same, seen from the left side.

Either spermatheca may open by two apertures, as did the hinder one in the specimen figured. × 2.

FIG. VI.—Two setæ, drawn to the same scale.

The right hand one is from the genital region, and the left from a postgenital segment.
* Indicates an axial differentiation. A. 3.

FIG. VII.—A small portion of the cuticle, with the cuticular sheath of a seta.
* These refractive dots appear to be due to the presence of enclosures. D. 3.

FIG. VIII.—The alimentary canal exposed from the dorsal aspect, its roof having been subsequently removed on the right side. × 2½.

The two hinder pairs of calciferous glands may often be absent, and the mesenteric septa in the crop-gizzard region are subject to variation.†

---

* In this and the following plate, the numbers of important segments are indicated wherever desirable in small Arabic numerals.

† The brownish-yellow tissue, usually termed liver, has no direct connection with the lumen of the alimentary canal. It is always associated very largely with the blood-vessels, and is in all probability a direct derivative of their walls. It appears to be active in the production of some constituent of the blood, similarly to that tissue described by Lankester in the Leeches as *rasifactive*. See Lankester "On the Vasifactive and Connective Tissues of the Medicinal Leech," Q. J. M. S., vol. xx., 1880. Also papers by Weldon on the Vertebrate Kidney and Supra-renal bodies, the same journal, vols. xxiv. and xxv. I have observed, in cells of the alimentary epithelium teased up shortly after death, the presence of *ingested* particles of decomposing vegetable matter. In the absence of true digestive glands, and of any knowledge of the physiology of alimentation in this animal, the probability of an intracellular digestion of this solidly ingested food material must not be overlooked. (Compare Hydra, Pl. xvii., fig. v., *et seq.*)

Fig. IX.—Enlarged drawing of the left half of somite XV., after removal of the crop specially to show the position and segments of the excretory organ. The loop marked * is ciliated. × 25.

In order to make out the natural relations of the excretory organ, it is advisable to preserve the animal in alcohol for a day or two, prior to dissecting.

Fig. X.—Terminal left-hand portion of Fig. IX. magnified, the looped blood-vessels of one side being indicated. D. 3.

These vessels usually bear numerous knotted appendages [see Lankester (48) and Claparede (42)]. They are often, as in the specimen here figured, wholly absent. D. 3.

Fig. XI.—A portion of the same, viewed in optical section. D. 3.

Fig. XII.—The mouth of the excretory organ. D. 3.

This can only be removed from the body, together with a portion of the mesentery which it perforates.

The whole structure of the Earthworm's excretory organ has been most successfully worked out by Gegenbaur (44), and Claparede (42).

Fig. XIII.—Dissection from the left side, showing especially the alimentary canal in section, and incidentally, the nervous system and genital organs. × 2½.

Eighteen hours' previous immersion in alcohol is here necessary, in order that the parts may be well set in their natural positions.

Of the œsophageal glands the two hinder pairs are always smaller than the one in front, and their apertures are sometimes obliterated (see Fig. VIII.)

PLATE XII

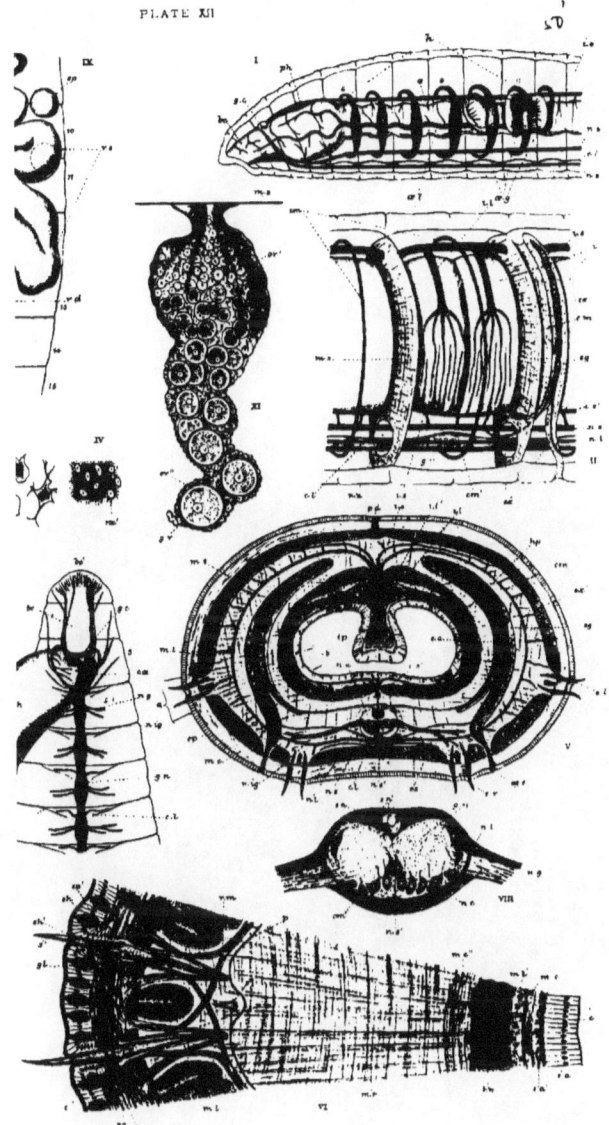

PLATE XII.

THE EARTHWORM.—The Organs of Circulation, Reproduction and Locomotion.
The Nervous System. The Layers of the Body-wall, etc.

Fig. I.—The so-called hearts, and the great blood-vessels of the anterior twelve segments, seen from the left side. × 2½.

Dissections of the blood system of this animal are best performed on specimens some six to eight hours dead.

Fig. II.—The vessels of three postgenital segments, seen from the left side.

The excretory vessels are shown in the posterior one, all except the excretory vessels in the middle one, and the circular commissural vessel in the anterior one.

In order to see the lateral-intestinal trunks, $i.l.$, the so-called hepatic tissue must be carefully scraped away.

The subintestinal vessel, $i.s'.$, Figs. II. and V., is very obvious on the under surface of the gizzard, the walls of which are white. × 6.

Fig. III.—A drop of the perivisceral lymph fluid, examined fresh.

The nuclei respond most readily to the action of 1 p.c. acetic acid and magenta. D. 3.

Fig. IV.—A drop of the red blood fluid, drawn from one of the so-called hearts with a pipette. F. 3.

Fig. V.—Entire section across the middle of the body.
Constructed from a number of sections and dissections.
The excretory organs are diagrammatic. × 6.

Fig. VI.—The portion a-b of the same, highly magnified, to show the order of succession of the layers of the body-wall and of that of the alimentary canal; also the setæ, with their sheaths, muscles, etc. D. 2.

The excretory organ is here disregarded.

Mojsisovics (50) has described and figured an epithelioid blood-plexus for the clitellum.

Fig. VII.—The anterior portion of the nervous system, seen from above.

The pharynx was displaced, and the left circumœsophageal commissure is supposed to be seen through it. × 3½.

Fig. VIII.—A portion of Fig. VI., being a thin section of the nerve ganglion with its muscular sheath, etc.

The nerve-cells here figured are diffused along the entire length of the nervous axis.
[Compare Claparède (42), Plate XLVII., Fig. VI.] D. 3.

The sections from which Figs. V., VI., and VIII. were drawn were prepared from an animal preserved in ½ p.c. chromic acid (the contents of its alimentary canal having been first washed out with the same fluid), and afterwards hardened in alcohol.

Fig. IX.—The reproductive organs, displayed from above.

The left half of the figure represents the immature, and the right half the sexually mature condition, so far as the male organs are concerned. Spirit specimen. × 6.

The spermathecæ generally underlie the mesenteric septa 9-10 and 10-11; they may, however, grow either into the ninth and tenth, or the tenth and eleventh segments.

Fig. X.—The initial stages in the development of the spermatozoa, obtained by squeezing out a drop of the contents of the seminal vesicle upon a slide. Stained with magenta. F. 3.

The figures d and f are drawn in optical section, e is seen *en face*, g represents three adult spermatozooids.

For further details concerning the male organs see Hering (45) and Bloomfield (41).

Fig. XI.—The ovary.

The ovum, $ov.''$, was ripe and ready for dehiscence. D. 3.

THE SNAIL.

PLATES XIII., XIV.

PLATES XIII., XIV.
THE SNAIL.

a.	Anus.	*h.gl.*	Ovotestis (hermaphrodite gland).
a.gl.	Albumen gland.	*h.gl'.*	Depression for the same.
a.gl'.	Duct of the same.	*h.j.*	Horny jaw.
al.	Archenteron.	*hy.*	Hypoblast.
ao.	Aorta.	*i.*	Intestine.
ao'.	Anterior aortic trunk.	*i'.*	Origin of the same.
ao".	Posterior aortic trunk.	*i.m.*	Intrinsic muscles.
au.	Auricle.	*l'.*	Circular lip.
aud.	Auditory organ.	*l".*	Lateral lip.
aud'.	Auditory nerve.	*lb.*	Labial nerve.
b.	Buccal nerve.	*l.m.*	Levator muscle.
b'.	Buccal ganglion.	*l.s.*	Lateral blood sinus.
b.d'.	Right bile-duct.	*l.t.*	Lateral (rachidian) teeth.
b.d".	Left bile-duct.	*lr.*	So-called liver.
bl.	Blastopore.	*lr'.*	Liver, right lobe.
b.m.	Buccal mass.	*lr".*	Liver, left lobe.
c.	Columella.	*m.*	Mouth.
c.c.	Cleavage cavity.	*m.gl.*	Accessory glands.
c.g.	Cerebral ganglion.	*ms.*	Mesosoma.
cl.	Genital vestibule.	*m.t.*	Median teeth (uncini).
c.m.	Constrictor muscle.	*nc.*	Nucleus.
c.p.	Cerebro-pedal commissure.	*n.c.*	Nerve collar.
c.ps.	Cerebro-splanchnic commissure.	*n.t.*	Nerve to tentacle.
cr.	Crop.	*n.t'.*	Nerve to optic tentacle.
cr'.	Its cut edge.	*o.c.*	Odontophoral cartilage.
d.	Dart sac.	*od.*	Oviduct.
d'.	Valves of the same.	*œ.*	Œsophagus.
d".	Spiculum amoris.	*op.*	Visual organ.
d.m.	Depressor muscles.	*op'.*	Optic nerve.
ep.	Epiblast.	*ot.*	Otolithic mass.
f.	Foot.	*p.*	Peritreme.
fl.	Flagellum.	*pc.*	Pericardium.
g.a.	Genital aperture.	*pd.*	Pedal ganglion.
g.d.	Genital duct.	*pd'.*	Pedal nerves.
gr.	Groove at base of excretory orifice.	*p.gl.*	Pedal gland.
g.s.	Germinal spot.	*pl.*	Pulmonary sac.
g.t.	Ganglion of so-called olfactory tentacle.	*pl'.*	Pulmonary chamber.
g.t'.	Ganglion of optic tentacle.	*pl".*	Respiratory aperture.
g.v.	Germinal vesicle.	*pl.m.*	Pallial muscle.
h.	Heart.	*p.m.*	Protractor muscle.
h.d.	Hermaphrodite duct.	*pn.*	Penis.
h.d'.	Aperture of the same.	*p.n.*	Pallial nerve.

pr.	Prostate.	s.gl.	Shell gland.
p.s.	So-called Parieto-splanchnic ganglion.	sl.	Salivary gland.
p.v.	Afferent pulmonary vessels.	sl'.	Salivary duct.
p.v'.	Efferent pulmonary vessels.	sl''.	Aperture of the same.
r.	Rectum.	s.n.	Nerves to the same.
r'.	Cut edge of the same.	sp.	Spermatheca.
rd.	Radula.	st.	Stomach.
rd'.	Sac of the radula.	t.	So-called olfactory tentacle.
re.	Renal organ.	t'.	Optic tentacle.
rd'.	Renal duct.	t.a.	Tentacular artery.
re''.	Renal aperture.	t.v.h.	Intestinal valve.
r.m.	Columellar (retractor) muscles.	v.	Subtentacular lobe (modified remnant of velum). For Plate XIII., ventricle.
r.m'.	Retractor pedis muscle.		
r.m''.	Retractor muscle of buccal mass.	va.	Auriculo-ventricular valves.
r.p.	Retractor muscle of penis.	v.c.	Blood sinus of visceral sac.
r.t.	Retractor muscle of tentacle.	v.c.p.	Circulus venosus pulmonis.
r.v.	Afferent renal vessels.	v.d.	Vas deferens.
r.v'.	Efferent renal vessels.	vg.	Vagina.
s.	Shell.	v.n.	Visceral nerve.
s'.	Its cut edge.	v.s.	Visceral sac.
sd.	Stomodæum.	v.s'.	Its cut edge.

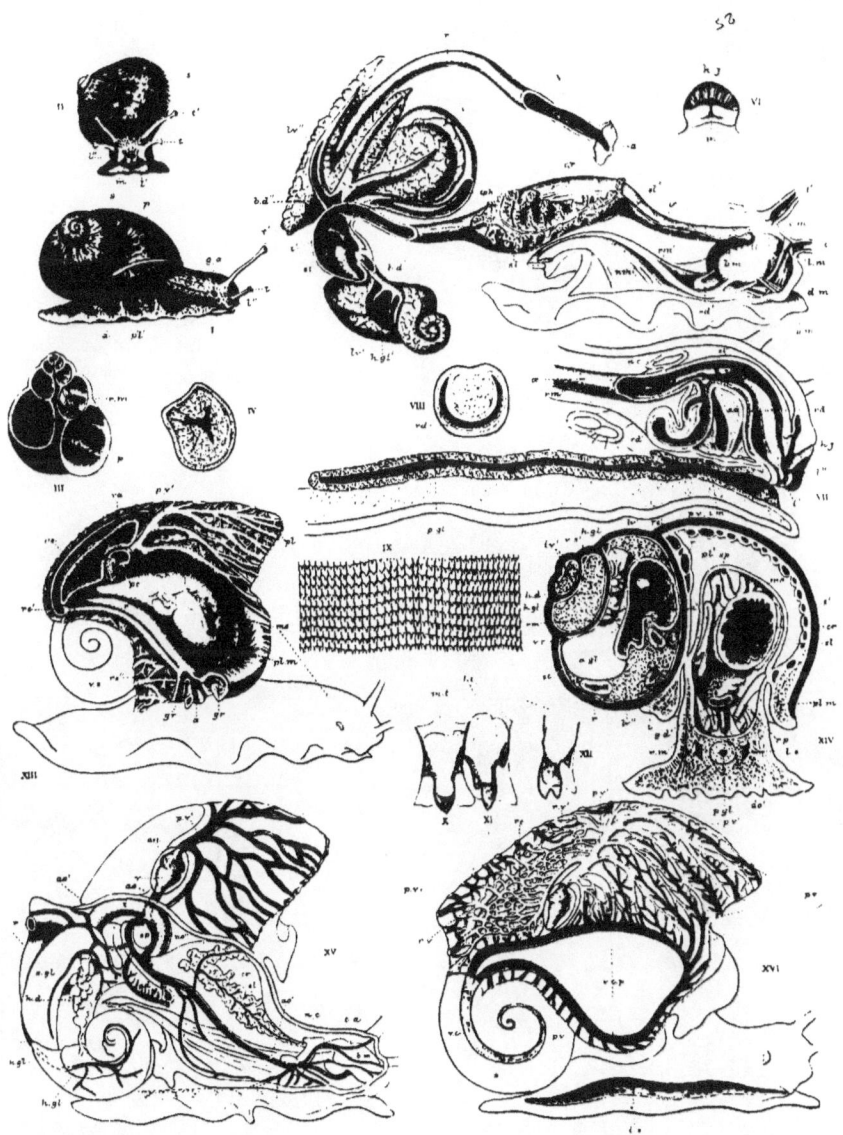

PLATE XIII.

THE SNAIL (*Helix aspersa*).—EXTERNAL CHARACTERS. THE SHELL. THE ALIMENTARY, CIRCULATORY, RESPIRATORY, AND EXCRETORY ORGANS.

FIG. I.—The living animal, from the right side.

FIG. II.—The same, from the front.

FIG. III.—The shell, in median longitudinal section through the columellar axis. The attached ends of the great retractor muscles were left.

FIG. IV.—The hybernaculum (hypophragm), after removal from the shell.

(Figs. I. to IV. all nat. size.)

FIG. V.—The alimentary canal, with its appended glands and associated muscles, from the right side.

On cutting open the pulmonary chamber, dissecting out the rectum and removing the generative organs and nervous system, the parts fall naturally into the positions here figured.

The stomach and portions of the bile-ducts and intestine have been opened up. × 2.

The so-called liver of the gasteropod is stated by Barfurth (56) to perform, among other functions, those of a hepato-pancreas. (Compare *Astacus*, Plate IX., Fig. VI.)

FIG. VI.—The horny jaw, viewed from the front in situ. × 3.

FIG. VII.—Median longitudinal section of the buccal mass and pedal gland, from the right side. × 4.

Sochaczewer (80) claims for the above gland an olfactory function.

FIG. VIII.—Transverse section across the sac of the radula. × 8.

FIG. IX.—A portion of the radula, magnified.
The middle row of teeth has been stippled. A. 2.

FIGS. X. and XI.—The median and first lateral tooth of the right side, in situ.

FIG. XII.—One of the outer lateral teeth of the same side. D. 2.

FIG. XIII.—Dissection to show the heart, pericardium, and the excretory organ.

The pulmonary sac was slit open in a line with the respiratory aperture, and the outer wall of the excretory organ removed, to show its internal structure. Both the pericardium and heart were opened up. × $2\frac{1}{2}$.

ATLAS OF BIOLOGY.

The excretory groove, *gr.*, cut across, lies in life altogether to the left of the respirator orifice.

The reno-pericardial communication is not figured; it is very small, and differs in no important respect from that drawn and described by Nüsslin (76) for *Helix pomatia*.

FIG. XIV.—Obliquely transverse section through the whole body and shell, taken just in front of the columella. Spirit preparation. × 3.

FIG. XV.—Dissection from the right side, after injection from the large pulmonary vein, *p.v.'*, to show the chief arteries.

The visceral sac and body-wall of the right side were dissected away *in toto;* the retractor muscles, genital ducts, and rectum were all in part removed, as shown by their cut ends drawn. The rectum was turned back, and the heart and right lobe of the liver a little displaced. All else is figured in situ. × 2.

The artery running longitudinally along the foot overlies the pedal gland.

FIG. XVI.—Dissection, after injection, to show the leading venous sinuses, together with the respiratory and renal capillary systems.

The arteries were injected from the pulmonary vein (afferent pulmonary vessel), *p.v.'*, and the venous lacunæ, pulmonary circulus, and afferent pulmonary vessels from the point marked *.

The pulmonary sac was severed close alongside the rectum, in order to show the whole respiratory plexus in one view. × 2.

The position of the lateral pedal sinus, *l.s.*, is indicated in life by a light band which runs along the side of the foot. On withdrawal of the body into the shell this sinus becomes greatly distended by displacement of the perivisceral fluid, and I know of no more satisfactory means of demonstrating a lacunar blood system, than that afforded by injection from this receptacle.

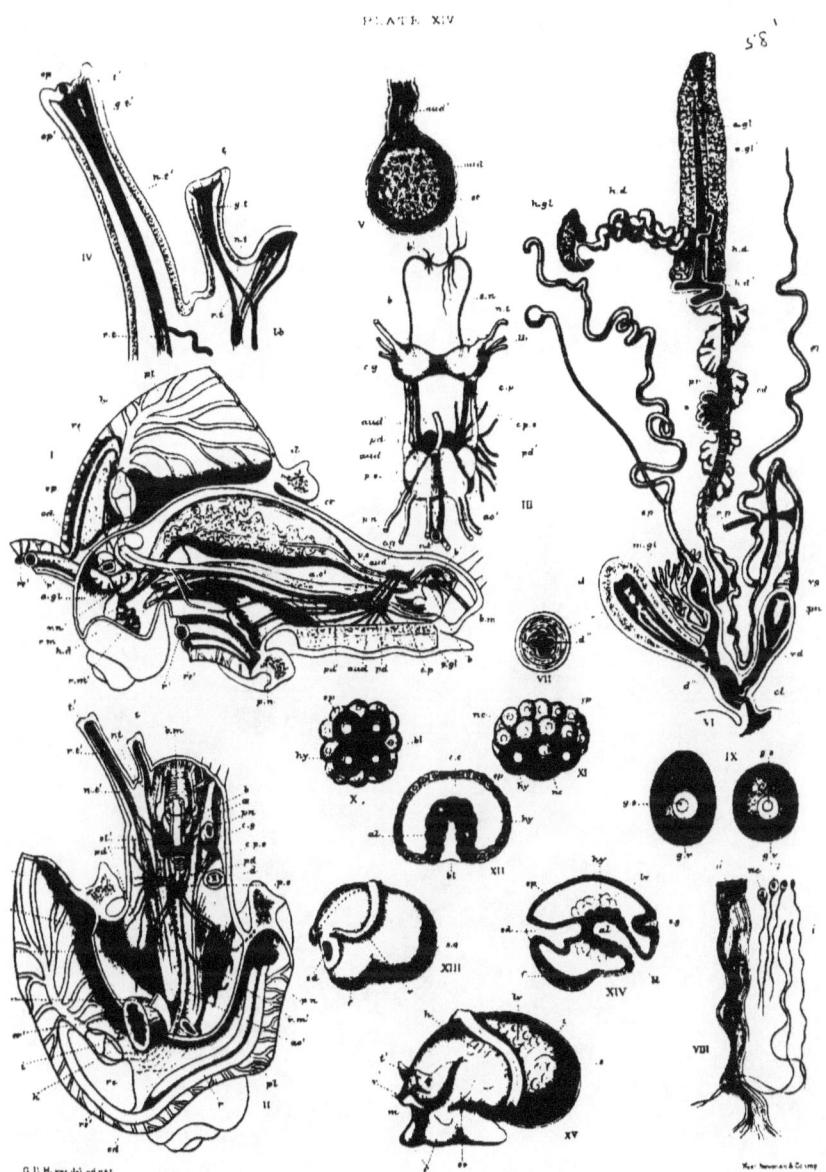

PLATE XIV

PLATE XIV.

THE SNAIL.—THE NERVOUS SYSTEM AND SENSE ORGANS. THE REPRODUCTIVE ORGANS AND THE GASTEROPOD LARVA.

FIG. I.—Dissection to show the leading ganglia and nerves, from the right side.

The lung-sac was opened up, cutting along the edge of the rectum, the rectum itself being next severed and reflected, as drawn. The body-wall was then removed on the right side from the head to the albumen gland, and along with it the tentacles, a good portion of the foot and the greater part of the genital apparatus. × 2.

FIG. II.—Dissection from above, to show some of the important nerves, and incidentally the tentacular and other retractor muscles.

The pulmonary sac and body cavity were opened up from above by a clean cut, passing to the right side of the excretory organ, and the two halves were reflected.

The tentacles of the left side were cut open. The crop, retractor muscle of the buccal mass, and the genital apparatus, were all for the most part removed, their cut ends being drawn. × 2.

In both the above figures the whole nervous system is drawn in deep black. The sheath of the nerve-collar, having been for the most part removed, is not indicated.

FIG. III.—The circumœsophageal nerve-collar, after removal from the body, seen from behind. × 3.

The dotted lines indicate the limits of its sheath.

The so-called parieto-splanchnic ganglia, *p.s.*, represent those known as visceral, pleural, and abdominal, in allied forms. Spengel (81) has recently instituted an elaborate inquiry into the whole question of their morphology.

FIG. IV.—An enlarged view of the tentacles of Fig. II., to show their nerves, ganglia, and the visual organ, in situ.† × 6.

FIG. V.—The otocyst, seen in situ. D. 3.

FIG. VI.—The generative apparatus, after removal from the body.

The vestibule, *cl.*, and adjacent parts have been opened up, at * is exposed the fold which incompletely subdivides the lower part of the hermaphrodite duct. × 2.

The ovotestis is best got at by removing it together with the right lobe of the liver, and then dissecting it out carefully. (Compare Figs. V. and XV., Plate XIII.)

---

† The condition of the nervous structures lodged within the optic tentacle clearly points to the conclusion that it performs a double function. See W. Flemming, "Untersuchungen über Sinnerepithelien der Mollusken," *Archv. Mk. Anat.*, vol. vi., 1870.

Fig. VII.—The dart-sac in transverse section. × 2.

Careful examination shows that the blades of the dart are slightly twisted, so that it must leave its sac in the fashion of a revolving gun-shot.

Fig. VIII.—The spermatozoa. F. 3.
i. Two isolated spermatozoa.
ii. A spermatozoan aggregate.
iii. A group of immature spermatozooids.

Fig. IX.—Two ova, obtained, as were the above spermatozoa, by teasing up a small piece of the ovotestis in eosin solution. D. 3.

Fig. X.—The ovum of the Pond snail* (*Lymnæus stagnalis*) during segmentation, seen from beneath. D. 2.

Fig. XI.—Side-view of the same.

Fig. XII.—The same, at the gastrula stage, in optical section. D. 2.

Fig. XIII.—The early larva of the same. Surface view from the side. D. 2.

Fig. XIV.—The above in optical section.

Fig. XV.—A twelve days' embryo of the same, from the left side. D. 2.

All from life, after treatment with 1 p.c. osmic acid. Figs. X. to XIII. after Lankester (72). The cilia are exaggerated.

---

* This type is chosen on account of the facility with which its larvæ can be obtained.

THE MUSSEL.

PLATES XV., XVI.

PLATES XV., XVI.
THE MUSSEL.

a,	Anus.		*lp".*	Nerve to the same.
a.a.	Anterior adductor muscle.*		*lv.*	"Liver."
a.a'.	Nerve to the same.		*m.*	Mouth.
a.ao.	Anterior aorta.		*mc.*	Micropyle.
al.	Albuminous fluid.		*na.*	Nacreous layer.
a,m.	Adductor muscle.		*nc.*	Nucleus.
a.p.	Posterior adductor muscle.*		*œ.*	Œsophagus.
a.p'.	Nerve to the same.		*p.a'.*	Right anterior pallial artery.
au'.	Right auricle.		*p.a".*	Left posterior pallial artery.
au".	Left auricle.		*p.ao.*	Posterior aorta.
b.d.	Bile-duct.		*p.ao'.*	Posterior aorta of right side.
b.d'.	Aperture of bile-duct.		*p.ao".*	Posterior aorta of left side.
b.n.	Branchial nerve.		*pc.*	Pericardium.
br.	Branchiæ.		*pd.*	Pedal ganglion.
br.a.	Afferent branchial vessel.		*pd.a.*	Pedal branch of anterior aorta.
br.a'.	Afferent branchial trunk.		*pl.*	Cells of embryonic mantle.
br.e.	Efferent branchial vessel.		*pl'.*	Right pallial lobe.
br.e'.	Efferent branchial trunk.		*p.m.*	Pallial muscle.*
br.l'.	Left outer gill lamina.		*p.n.*	Pallial nerve.
br.l".	Left inner gill lamina.		*po.*	Periostracum.
by.	Byssus.		*p.p.*	Protractor pedis muscle.*
by'.	Byssus gland.		*pr.*	Prismatic layer.
c.c.	Supraœsophageal commissure.		*p.s.*	So-called parieto-splanchnic ganglia.
c.g.	Cerebral ganglion.		*p.r.*	Efferent pallial vessels.
c.p.	Cerebro-pedal commissure.		*r.*	Rectum.
c.p.s.	Cerebro-splanchnic commissure.		*r'.*	Its cut edge.
e.s.	Exhalent siphon.		*re.*	Glandular portion of excretory organ.
f.	Foot.		*re'.*	Nonglandular vestibule of excretory organ.
g.a.	Genital aperture.		*re".*	Renal aperture (or style passed into the same).
g.d.	Genital duct.			
g.gl.	Genital gland.		*re.c.*	Interrenal aperture.
g.s.	Germinal spots.		*re.p.*	Reno-pericardial aperture.
g.v.	Germinal vesicle.		*r.m'.*	Anterior retractor muscle.*
i.	Intestine.		*r.m".*	Posterior retractor muscle.*
i.b.c.	Infrabranchial chamber.		*r.m'".*	Lesser retractor muscle.*
i.s.	Inhalent siphon.		*r.v.*	Afferent renal vessel.
kb.	Keber's organ.		*r.v'.*	Efferent renal vessel.
lg.	Ligament.		*s.*	Valves of shell.
lp.	Labial palps.		*s'.*	Inturned edge of shell.
lp'.	Artery to the same.		*s.b.c.*	Suprabranchial chamber.

Note.—The references marked thus, *, refer in Fig. III. to the crests of attachment of these muscles.

s.n.	Siphonal nerve.	va'.	Right auriculo-ventricular valve.
st.	Stomach.	va".	Left auriculo-ventricular valve.
tc.	Tactile organs of larva.	v.c.	Central blood sinus.
ty.	Typhlosole.	v.m.	Vitelline membrane.
u.	Umbo.	vs.	Visceral branch of anterior aorta.
v.	Ventricle.		

PLATE XV

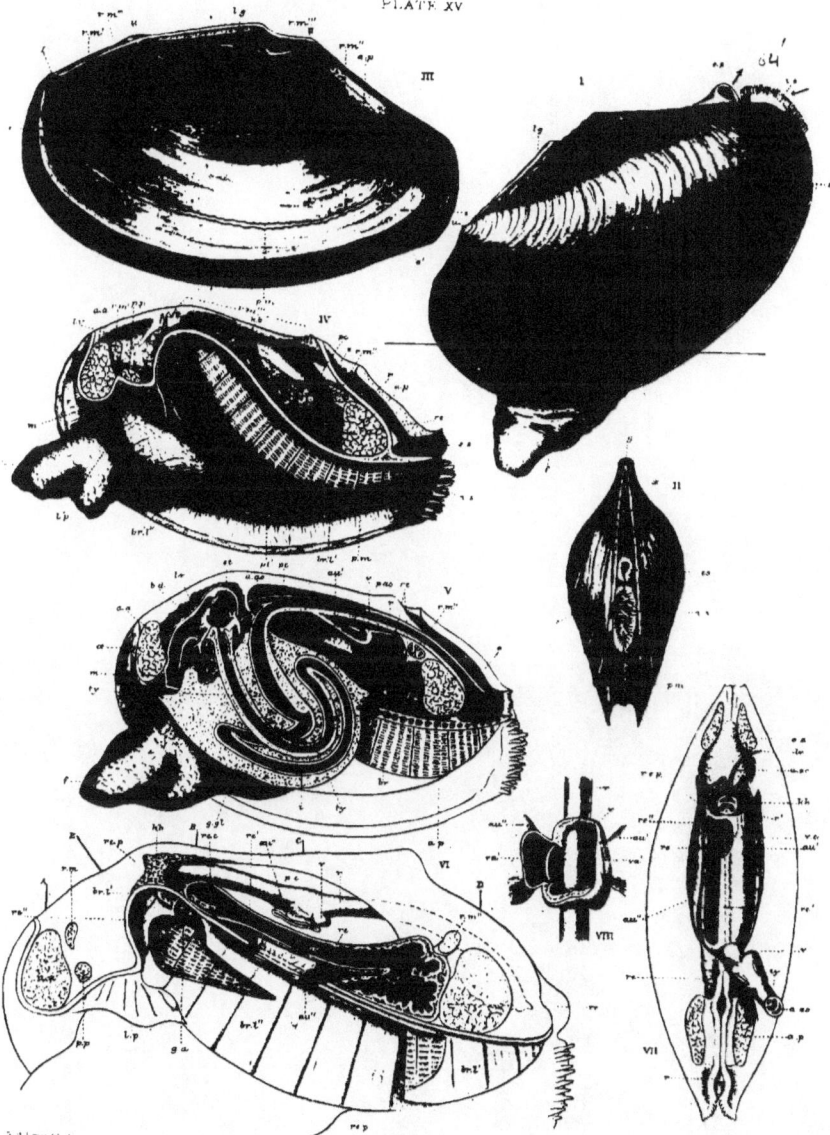

PLATE XV.

THE FRESH-WATER MUSSEL.—THE EXTERNAL CHARACTERS AND THE SHELL. THE ALIMENTARY AND EXCRETORY ORGANS. THE HEART, ETC.

FIG. I.—The living animal drawn from the left side, as seen in a deep, laterally compressed glass vessel half filled with sand, the surface level of which is represented by a transverse line.

The arrows indicate the course of the induced ciliary currents.

FIG. II.—The same, seen from behind.*

FIG. III.—The right valve, from within.

If carefully dissected from the body under water, it will be seen that the superficial chitinous layer is continuous on the hæmal side, between the points ɪ. ɪɪ., and also for the area marked s'., where it is turned in and reflected on to the pallial muscle. (Compare the transverse sections figured on Plate XVI.)

FIG. IV.—Dissection from the side, the left valve and pallial lobe alone removed.

The cut edge of the mantle is specially shown, as it indicates the line of attachment of the gills and labial palps, and consequently that of demarcation between the supra and infrabranchial chambers.

In the specimen figured the external gill lamina was fully distended with embryos.

FIG. V.—The same, dissected to the level of the alimentary canal;† this, the pericardium, heart, and aortæ, have all been opened up.

The labial palps were cut down to the level of the middle line.

In this figure the suprabranchial chamber, incompletely subdivided by the suspensory ligament of the branchiæ, is also seen (compare Figs. VII. and VIII. of Plate XVI.)

(Figs. I. to V. all nat. size.)

FIG. VI.—The excretory organ, dissected from the left side, shortly after death.

The left half of the mantle-lobe was first removed as for Fig. IV., and after that the greater part of the outer, and a small portion of the outer wall of the inner gill laminæ,—as indicated by their cut ends drawn. The genital and excretory apertures being thus exposed,

---

* The larger tentacles of the inhalant siphon are undoubtedly sensory, and I have witnessed the ejection of embryos through the pore marked *, first figured and described by Kober (70). Compare Figs. IV. and V.

† If any difficulty is experienced in following the coils of the alimentary canal, it can be overcome by first injecting with plaster of Paris. This method should, however, be used with caution, as the parts are liable to be unnaturally distended.

the left half of the pericardium was removed, and the auricle of that side reflected. The excretory organ was then opened up from the side, and its contained excreta gently washed away. × 1½.

The fold seen embracing the genital and excretory orifices is derived from the lower lip of the genital aperture.

The mussel is at all times best dissected in its shell, but if otherwise, pins should be preferably passed through the adductor muscles. If, with the animal thus quite rigid, the above directions are closely followed, there can be no difficulty in following the relations of the excretory organ, provided it is clearly remembered that that organ is altogether outside and below the pericardium. (Compare Fig. VII., and also Figs. VI. and VII. of Plate XVI.)

FIG. VII.—Dissection of the pericardium and excretory organ from above.

The roof of the pericardium was cut away, the auricles in part removed, and the rectum reflected. On the left side the floor of the pericardium and the roof of the excretory vestibule, re'. (confluent in life), have been removed; the point at which this is most obvious indicates the interrenal aperture.

The limits and relations of the organ of Keber are shown, and the anterior aorta has been opened up on the right side.* Nat. size.

FIG. VIII.—The heart in situ, laid open from above.

The greater part of the roof of the ventricle and of the auricle on the left side have been removed, to the level of the auriculo-ventricular valve. × 3.

The student must not be misguided by the apparent anomaly in the passage of the alimentary canal through the heart; for be it remembered that in most animals, at one period or other of their existence, the two things are closely related.

---

* I cannot satisfy myself that the supposed communications between the blood-vascular system and the pericardium —first described by Keber (70)—have any real existence. They appear to me to have been the products of undue pressure used in injecting, and my own observations are entirely in harmony with those of Cattie (68) and Lankester (74), and opposed to those of Griesbach (65).

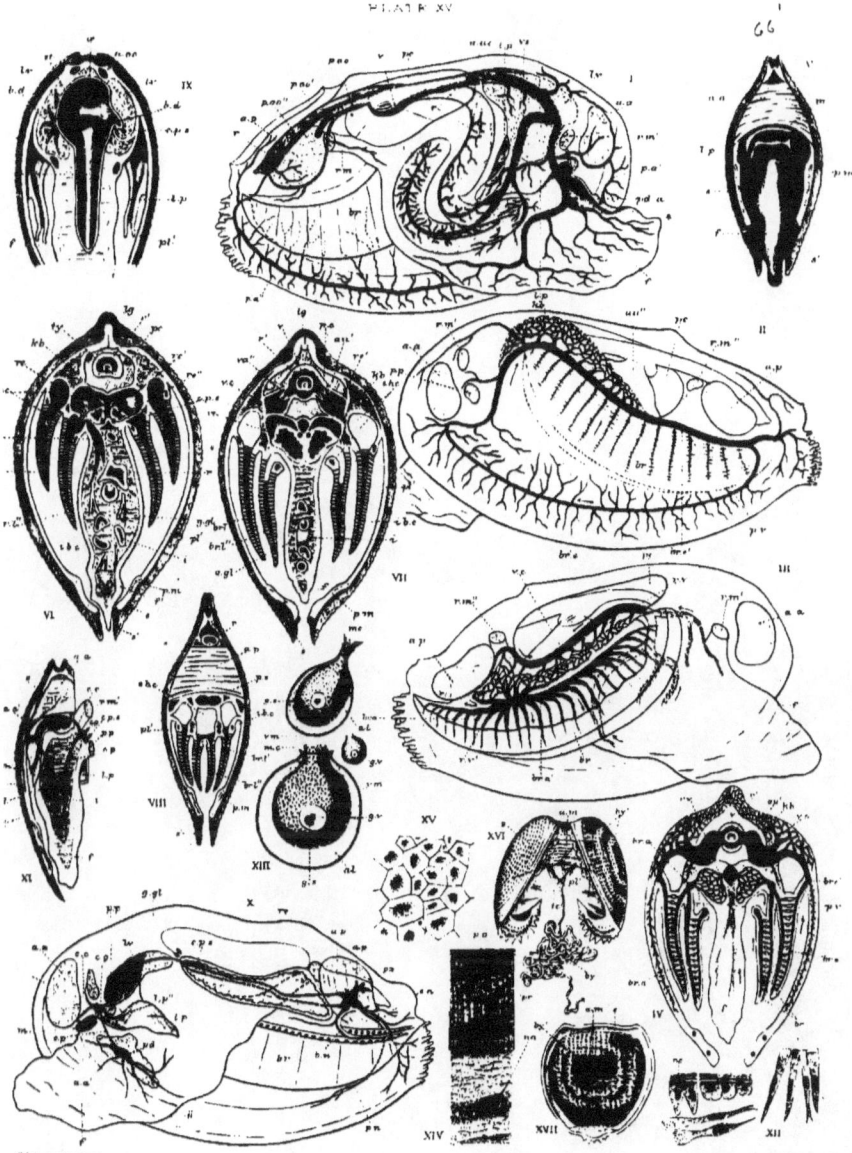

PLATE XVI.

THE MUSSEL.—Organs of Circulation. The Nervous System. The Embryo, etc.

Fig. I.—The arterial system, seen from the right side, after injection from the ventricle.
With the exception of a small anterior portion marked *, the right pallial lobe has been removed, and with it the gills of that side. The pericardium was opened up, and the mesosoma removed to the level of the alimentary canal. Nat. size.

Fig. II.—The efferent pallial, efferent branchial and related vessels, seen from the left side after injection from the auricle.
The efferent branchial trunks are drawn, as seen after removal of the mantle, and no note is taken of the extensive system of pallial sinuses connected with the trunk, $br.e'$. Nat. size.

Fig. III.—Dissection from the right side, to show the great veins in relation to the excretory organ.
The mantle lobe was removed, and the external gill lamina opened up.
The afferent branchial trunk and its branches, and the so-called vena cava were injected direct, after removal of the right auricle.
For further details see Langer (71).
Fleming (64) records some observations upon the blood corpuscles and lacunæ of this animal, together with an ingenious method of injection. Nat. size.

Fig. IV.—Slightly diagrammatic representation of a transverse section across the middle pericardial region, after injection. × 1½.
The channels shaded lightly, with the exception of that adjacent to the letter f., are those carrying the aerated blood. × 1½.

Fig. V.—Transverse section across A. of Fig. VI. of Plate XV., seen from the front, to show the relations of the adductor muscle to the valves, and of the labial palps to the mouth.

Fig. VI.—Transverse section across B. of the same, the excretory and genital ducts having been opened up on the left side.

Fig. VII.—Transverse section across C. of the same, passing through the auriculo-ventricular valves and the great aperture of communication between the supra- and infra-branchial chambers.

Fig. VIII.—Transverse section across D. of the same.

FIG. IX.—Oblique transverse section across E. of the same, to show the so-called liver, the main bile-ducts, and the relations of the lesser retractor muscles.

The above sections were made from a fresh specimen, some few hours after death. All × 1½.

FIG. X.—The nervous system, dissected from the left side, only such additional parts being indicated as are concerned in ascertaining its course.

The pallial lobe and gills were removed, the excretory organ of the left side was slit open, and its contents washed out.

The cerebral ganglia having next been found, the body-wall, liver, and genital gland were in part removed as indicated—cutting in a line with the tendon of the protractor pedis muscle, in order to expose the whole course of the cerebro-splanchnic commissure, *c.p.s.* Nat. size.

The so-called parieto-splanchnic ganglia of the *Lamellibranch* have been shown by Spengel (81) to represent the olfactory ganglia of other molluscs.

FIG. XI.—Oblique transverse section across the oral region of Fig. X., to show the relations of the supraœsophageal nerve-commissure, the pedal ganglia, and, on the left side, the whole course of the cerebro-pedal commissure.

The valve and labial palps of the left side were for the most part removed, the origins of the tendon of the anterior pedal muscles being shown incidentally. Nat. size.

FIG. XII.—Ciliated and other cells, scraped from the lining membrane of the intestine. The two lower ones were active in the production of a secretion. D. 3.

FIG. XIII.—Three ovarian ova, at different stages, teased out in eosin solution. D. 3.

The protoplasmic filament passing through the micropyle, *mc.*, represents the torn neck of attachment to the germinal epithelium. See Von Jehring (67).

FIG. XIV.—Transverse section of the valve, cut at right angles to its long axis. Prepared as described for the Crayfish shell (Plate X., Fig. VIII.) A. 3.

FIG. XV.—Tangential section across the prismatic layer of the same. D. 2.

FIG. XVI.—The *Glochidium* larva, from the gills of an animal killed under chloroform. Spirit material, stained with magenta.

FIG. XVII.—The same, seen from the left side.

The dots on the valve represent pore canals.

PLATE XVII.

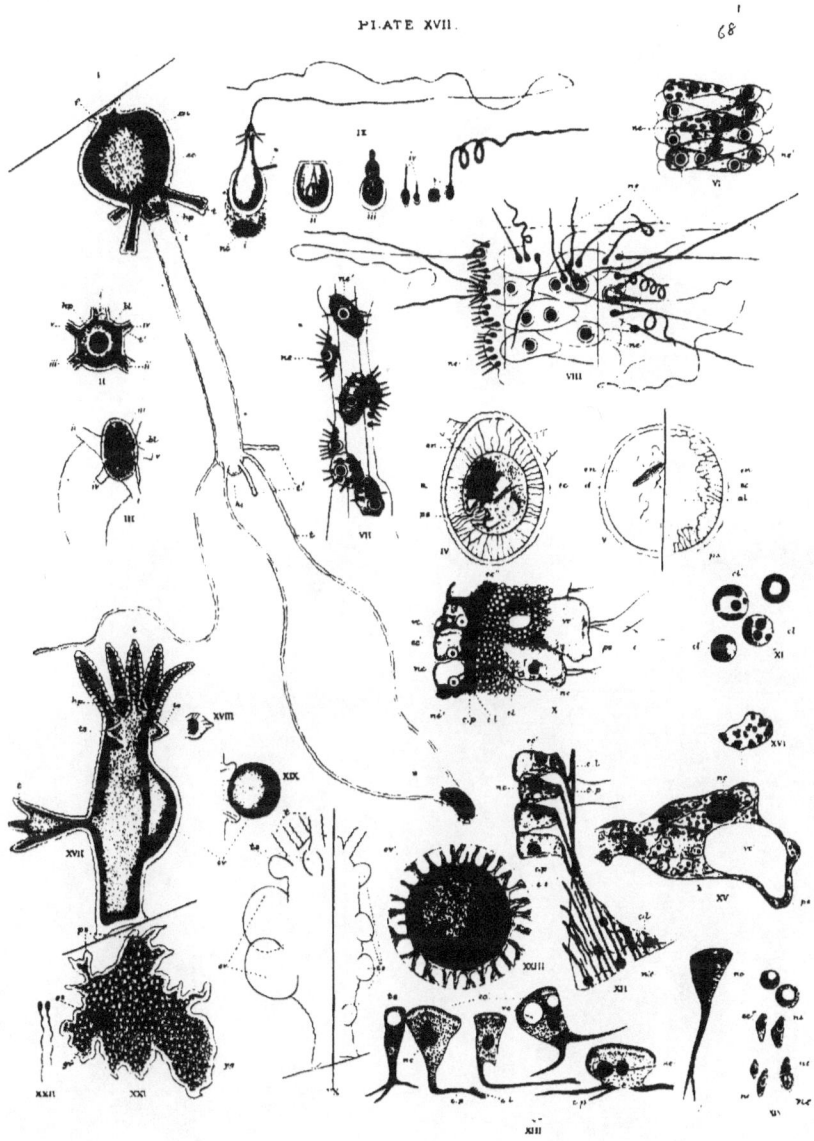

PLATE XVII.

THE HYDRA.

FIG. I.—The green Hydra, at the maximum of contraction and elongation of its body. The animal is drawn in the act of seizing a small Cypris, marked *. A. 2.

FIG. II.—The same, looked at from above.

FIG. III.—The above, with the hypostome everted.
In this specimen two young tentacles were just appearing, the numerals i. to v. indicate the order of development of these organs. Compare Jung (90).

FIG. IV.—Transverse section across the body of a Hydra, in the digestive cavity of which a small crustacean was present.
At * are seen the remnants of the digestible parts of that organism, not yet assimilated. *Hydra fusca.* A. 4.

FIG. V.—Outline sketches of portions of two sections across the body of the same animal, showing the variable extent to which the central cavity may be obliterated by the amœboid activity of the endoderm cells.
At d. a diatom is seen, bodily ingested. *H. fusca.* A. 4.
The most recent researches upon this intracellular digestion in the *Hydrozoa*, are those of Lankester (95). A paper on the subject generally, by Mentschnikoff, will be found in vol. xxiii. of the same journal.

FIG. VI.—A small portion of one of the tentacles, in the contracted condition. Surface view. D. 2.

FIG. VII.—The same in elongation.
In both these figures, only the nematocyst-bearing cell aggregates are shown. The small nematocysts, alone drawn in Fig. VII., are fixtures.

FIG. VIII.—The same, contracted under the influence of an irritant (1 p.c. acetic acid.)
The larger nematocysts are drawn as observed to become everted. They stain very readily with magenta solution. D. 3.

FIG. IX.—The leading types of thread-cells, drawn after liberation from the body.
i represents the functional, and ii the resting condition of the larger nematocysts; in i, the nucleated mass at the base and the fragment marked * represent the remains of the parent cell.

iii. A younger example of the same, everted.
iv. The small fixed nematocysts.
v. The resting and active conditions of a third variety. The filaments of these stain very deeply with magenta. All drawn to same scale. F. 3.

Hartog (87) has accounted for the occasional presence of thread-cells in the endoderm.

FIG. X.—Small portion of a transverse section across the body of a green Hydra. Picric acid, alcohol, borax carmine.

The cilia are drawn from an osmic acid preparation. (See Parker 97.) D. 3.

FIG. XI.—Isolated examples of the chlorophyll-bearing bodies of the same. Teased up fresh in water. Gundlach's $\frac{1}{16}$th immersion.

FIG. XII.—A small portion of a similar section to Fig. X., with a piece of the supporting lamella, c.l., seen *en face*.

Neither interstitial tissue nor endoderm cells are drawn. *H. fusca*. Osmic acid. D. 3.

FIG. XIII.—Larger cells of the ectoderm, isolated by bichromate of ammonia. F. 3.

FIG. XIV.—Cells of the interstitial tissue, treated in the same manner.*

FIG. XV.—The greater portion of a solitary endoderm cell from *H. fusca*, isolated. Picric acid, and borax carmine. F. 4.

FIG. XVI.—One of the sooty-particle bearing portions of the same. Gundlach's $\frac{1}{16}$th immersion.

Lankester (94) has suggested the most recent interpretation put upon these bodies, and has finally set at rest the real nature of the chlorophyll-bearing bodies of *H. viridis*, referred to above. The facts set forth in his paper have a most important bearing upon the probable unity of the green and brown Hydræ.

FIG. XVII.—A large brown Hydra, bearing at the same time asexually produced buds and sexual organs. A. 2.

FIG. XVIII. a testis, and XIX. an ovary of the above, at a later stage of development.

FIG. XX.—Outline sketches of portions of two brown Hydræ.

The left one bore three ovaries, two of which are figured. The right-hand one bore nine testes, of which four are figured.

This drawing does not by any means represent the maximum development of the male organs, but it suffices to show that the testes need not necessarily appear only at the bases of the tentacles.

* I am strongly of opinion that these often form syncytia.

FIG. XXI.—The ovum of Fig. XVII., liberated under gentle pressure. D. 2.

FIG. XXII.—Two ripe spermatozoa from the same. F. 4.

FIG. XXIII.—The ovum, at a late stage of segmentation, still enveloped in its membranes. D. 2.

The sexual reproduction of Hydra has long been a vexed question; the last contribution to the subject is that of Korotneff (92).

al.	Digestive (body) cavity.	g.v.	Germinal vesicle.
bl.	Mouth (blastopore).	hp.	Hypostome.
c.	Cilia.	uc.	Nucleus.
cl.	Chlorophyll-forming bodies.	ne.	Nematocysts (various).
cl'.	Chlorophyll corpuscles.	ne'.	Larger nematocysts, before rupture.
e.l.	Supporting lamella.	ov.	Ovary.
e.p.	Kleinenberg's fibres.	ov'.	Ovum.
d.	An ingested diatom.	ps.	Pseudopodia.
ec.	Ectoderm.	t.	Tentacles.
ec'.	Larger ectoderm cells.	t'.	Young (last formed) tentacles.
ec''.	Smaller ectoderm (interstitial) cells.	ts.	Testis.
en.	Endoderm.	vc.	Vacuole.
e.s.	Egg shell.	vc'.	Amyloid vacuole of endoderm cell.
f.	So-called foot.	y.g.	Yolk granules.
g.s.	Germinal spots.		

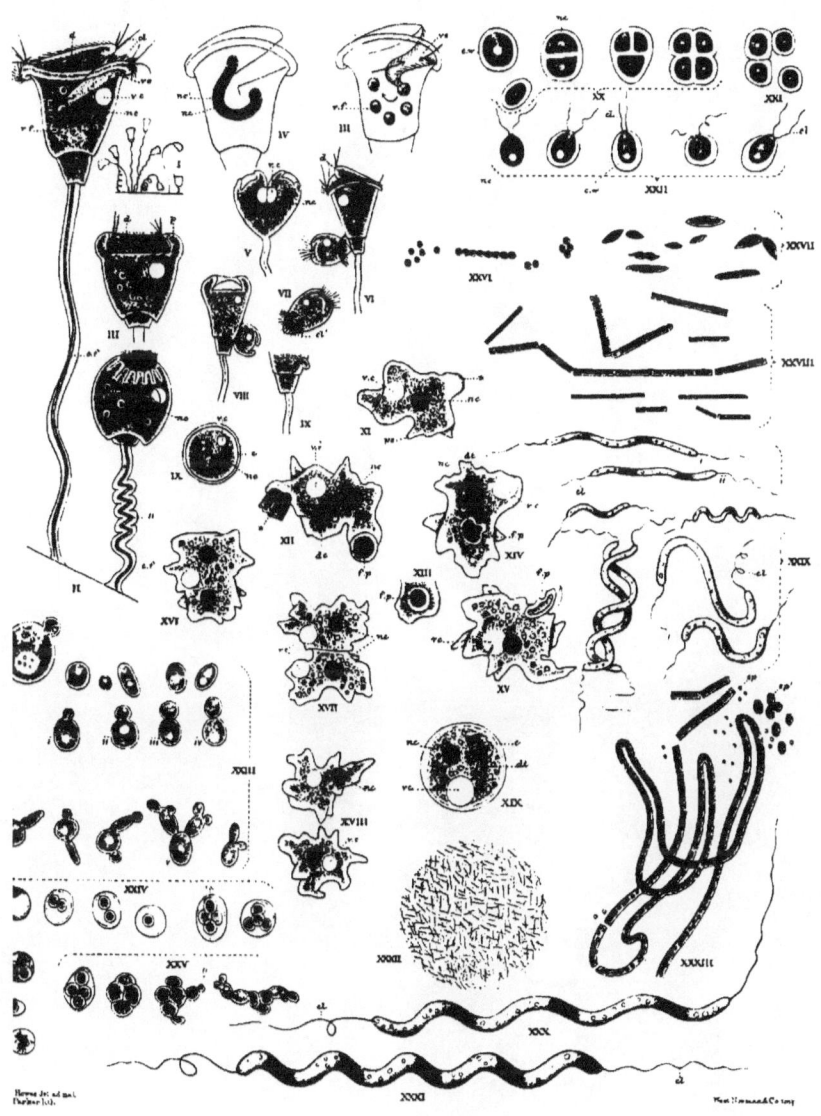

PLATE XVIII.

THE UNICELLULAR ORGANISMS.

Vorticella.

Fig. I.—A group of Vorticellæ, showing the animal in various positions. A. 3.
For a summary of the numerous species of this prolific genus see Kent (113).

Fig. II.—The same animal, in the extended and retracted states. Surface views.

Fig. III.—Another individual, drawn in optical section during the process of extension.

Fig. IV.—Two views of the same. In the right hand one (erroneously lettered Fig. III.) a series of food vacuoles are drawn, one of them in the act of being ingested.
The arrows indicate the course taken by these, and the top left-hand one marks the point at which egestion most frequently takes place.
The left-hand figure represents the nucleus, after treatment with acetic acid and magenta.
(Figs. I. to IV., all drawn under Zeiss. D. 4.)

Fig. V.—The initial phase in the process of multiplication by fission.
Two hours later, two equal-sized organisms resulted from this.

Fig. VI.—An example in which two individuals, unequal in size, resulted from the same process.
The smaller one was drawn in the act of liberating itself.

Fig. VII.—The same, after liberation.

Figs. VIII. and IX.—Two successive phases observed in the process of conjugation.

Fig. X. (Erroneously lettered IX.)—An encysted Vorticella.
(Figs. V. to X. all drawn under D. 2. The cilia are represented only in VI. and VII.)

Amœba.

Figs. XI. to XVIII. represent successive phases in the life-history of an Amœboid organism, kept under constant observation for three days. See Appendix I.

XI.—The locomotor phase.
At * the ectoplasm is seen in the act of protrusion to form a pseudopodium, the endoplasm passed into it later and quite suddenly.

XII.—A period in the ingestive phase.

The animal was observed to move round the glassy fragment figured *, for a period of twenty minutes, finally rejecting it in favour of the small vegetable organism, *f.p.*; in all probability, any nutritive matter which may have been adherent to it was meanwhile taken up.

XIII.—A portion of Fig. XII., after the organism, *f.p.*, had been ingested.

XIV. AND XV.—Successive stages in the assimilative and excretive processes; XV. was drawn 19—20 hours later than Fig. XIII. The digestible parts of the ingested organism were assimilated by the Amœba, and the refuse ejected, as figured, in a distorted disintegrating condition.*

The contractile vacuole, *c.v.*, is drawn in Fig. XV., at both systole and diastole.

Engelmann (103) has made some observations upon the physiology of this organ.

XVI., XVII., XVIII.—Successive stages observed in the reproductive process of the same organism, two days later. XVI. represents the nuclear-division stage, and the two remaining figures are phases in the division of the cell.

The organism figured above was a small Amœba having a conspicuous nucleus, found in the sediment from a small fresh-water aquarium. Its probable specific identity is left an open question, and the observations of Wallich (125) show how far such a determination might be reliable upon purely external appearances.

Gruber (107 to 110) has recently entered upon investigations of the greatest value, concerning both this subject and the behaviour of the nucleus among unicellular animals.

FIG. XIX.—A similar organism, encysted.

Three or four hours later the cyst disintegrated, liberating the animal.

(Figs. XI. to XIX. all drawn under Zeiss. D. 3.)

The Protococcus.

Figs. XX. to XXII. represent successive stages observed in the life-history of Protococci, scraped from the bark of a tree.

A somewhat similar difficulty arises here as with Amœba, with reference to the specific identity. The latest paper on the subject is that of Klebs (114).

XX.—A group of organisms in the dried state.

The three stages in division figured, were observed in one individual.

---

* The appearances presented were such as to presuppose a digestive action on cellulose.

THE UNICELLULAR ORGANISMS.

XXI.—One of the same, after two or three days' immersion in water under the microscope.

XXII.—Later phases in the motile stage assumed by the above.
The black spot indicates the red colouring-matter present, and the extreme left-hand figure is that of an individual destitute of cellulose investment. (Bergh, *Morph. Jahrb.*, vol. vii., 1881, has discovered a cellulose investment for certain chlorophyll-bearing *Protozoa*.)

The Yeast Plant.

Fig. XXIII.—Cells of ordinary brewer's yeast.
i.-iv. Stages in division of the same cell. D. 4.
i.* Fig i. as seen under Gundlach's $\frac{1}{15}$th immersion.
v. A branching colony, still retaining their original connections.

Schmitz, *Stzb. Bonn*, August, 1878, has described in the yeast-cell, and in certain other low organisms, Mucor among their number, what he considers to be a nucleus.

Fig. XXIV.—The endogonidia (ascospore) phase of reproduction, as seen in a sample of yeast sown on a slab of gypsum. F. 3.
See Huxley (111) and Rees (120).

Fig. XXV.—Further development of the endogonidia, after transfer to Pasteur's solution. F. 3.

The Bacteroid.

Fig. XXVI.—Micrococcus.

Fig. XXVII.—Bacterium.

Fig. XXVIII.—Bacillus.
The central filament of this series segmented up, as drawn, within ten minutes of its detection.

Fig. XXIX.—Spirillum.
i., ii. represent the so-called Vibrio. It is probably but a stage of Spirillum.

Figs. XXX. and XXXI.—Two giant Spirilla.

Fig. XXXII.—A drop of the surface scum, showing a Spirillum aggregate in the resting state.

Fig. XXXIII.—The so-called spore-forming stage in Spirillum.
The specimen segmented up as indicated, while being drawn.

Brefeld (99) gives figures of the germination of the so-called spores of Bacillus, drawn at recorded intervals of time.

All the above *Bacteroids* were observed in some hay-infusion, allowed to stand two days in a warm room. They are all—with the exception of Fig. XXXII.—drawn to the same relative scale, as viewed under Gundlach's $\frac{1}{15}$th immersion.

Dallinger (102) has demonstrated the presence of cilia in the motile stages of Bacterium and Bacillus.

b.	Bud.		nc.	Nucleus.
c.	Cyst.		nc'.	Nucleolus.
c.f.	Contractile fibre.		p.	Peristome.
cl.	Cilium.		ps.	Pseudopodium.
cl'.	Posterior ring of cilia.		sp.	So-called spores.
c.w.	Cell-wall.		sp'.	Encysted spore-like masses.
d.	Disc.		v.	Vacuole.
dt.	Ingested diatom.		c.v.	Contractile vacuole.
eg.	Endogonidia (ascospores).		v.f.	Food vacuole.
f.	Fat drops.		vs.	Vestibule.
f.p.	Food particle.			

PLATE XIX.

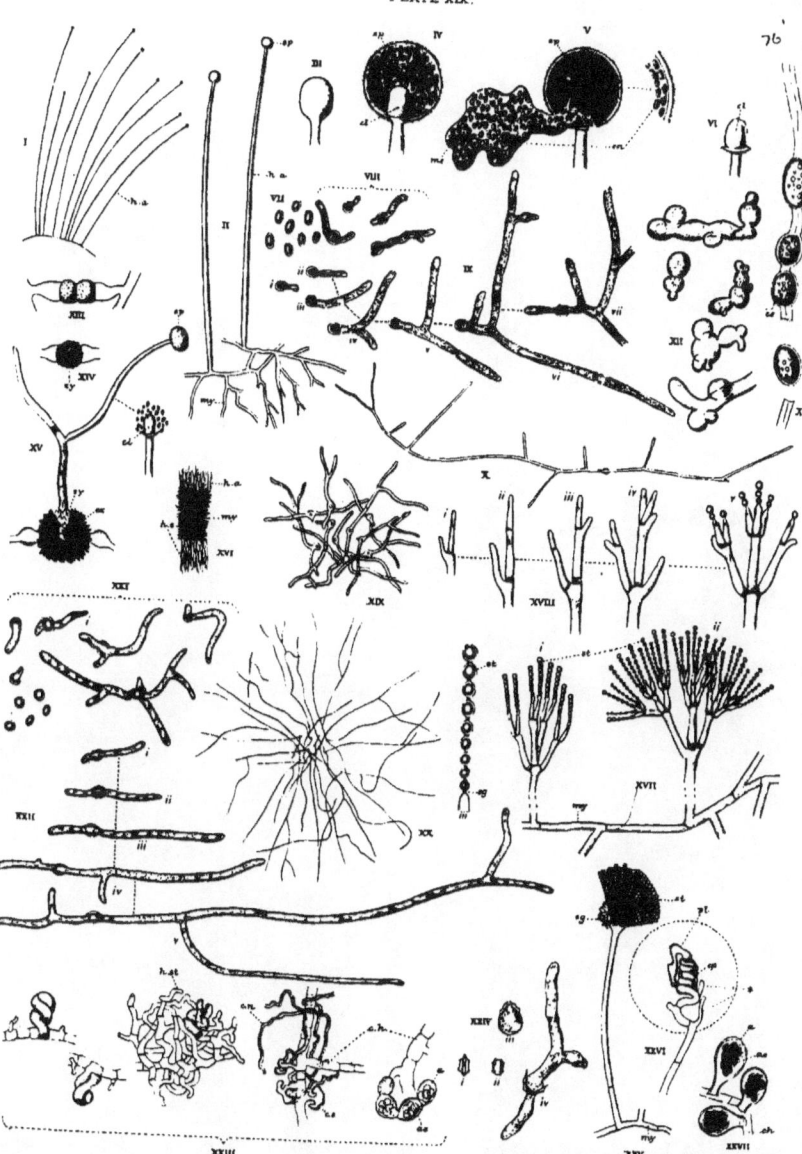

PLATE XIX.

THE FUNGI.

Mucor mucedo.

Fig. I.—Spore-bearing hyphæ of mucor, as seen in life, standing out from a piece of horse-dung.

Fig. II.—A small portion of the same fungus, teased out with needles. A. 4.

Figs. III., IV., V.—Successive stages in the development of the sporangium.
IV. is represented in optical section.
V. A ripe sporangium crushed under pressure (the mucilaginous fluid, *mc.*, is represented too darkly).
The small figure to the right is one of a portion of V. in optical section.

Fig. VI.—The central columella of the above.
(Figs. III. to VI. all D. 2.)

Fig. VII.—Isolated spores of mucor.

Fig. VIII.—Germinating spores of the same.

Fig. IX —i. to vii. successive stages in the germination of a single spore.
Only a portion of vii. is drawn.
Object glass culture, see Appendix I.
(Figs. VII. to IX. all D. 4.)

Fig. X.—The product of twenty-four hours' growth from a single spore, itself indicated as a slight enlargement.
Object glass culture. A. 3.

Fig. XI.—A chlamydospore—bearing hypha of the same. One spore is seen breaking away. D. 4.

Fig. XII.—The torula stage of Mucor, from a growth kept submerged in a saccharine solution. D. 4.
See Max Rees (120) and Huxley (111).

Figs. XIII., XIV., XV.—Successive phases in the conjugative process of Mucor.
From Brefeld (128).

Penicillium glaucum.

FIG. XVI.—A small portion of the crust of Penicillium in transverse section. A. 2.

FIG. XVII.—A piece of the above, teased up.
i., ii. Two adjacent conidia. For further varieties see Brefeld (129). D. 4.
iii. One series of stylo-gonidia of the same, more highly magnified. F. 3.

FIG. XVIII.—Successive stages observed during one day's (10 hours) growth of a conidiophore of Penicillium.
Object glass culture. D. 4.

FIG. XIX.—A group of germinating Penicillium spores.
Object glass culture. D. 2.

FIG. XX.—The same, sixty hours after sowing. A. 2.

FIG. XXI.—Isolated spores of this fungus.
Many are germinating. D. 4.

FIG. XXII.—Successive phases in the germination of a single spore. D. 4.

FIG. XXIII.—Phases in the conjugative reproduction of Penicillium.

FIG. XXIV.—Germination of one of the ascospores, produced as the result of conjugation.*
[Figs. XXIII. and XXIV. after Brefeld (129).]

Aspergillus.

FIG. XXV.—Spore-bearing hypha of this fungus, from a slice of bread exposed to a warm moist atmosphere for 15 days. D. 3.

FIG. XXVI.—The conjugative phase of the same.
Obtained by teasing up the yellow portions of the fungus, grown as above. The hyphæ marked * are two last developed ones of a great number (not drawn) which formed the investing peritheca, the limits of which are indicated by a dotted line. D. 4.

FIG. XXVII.—Portion of a spore-bearing carpogonial hypha of the above, at a later stage. F. 3.
[Figs. XXV. to XXVII. from life, after de Bary and Woronin (126).]

* The chances of the student's obtaining a cultivation of the conjugative phase of this fungus being but slight, the Eurotium stage of aspergillus is supplemented, as it can be readily obtained.

a.	Ascus.	*h.a.*	Aerial hyphæ.
as.	Ascospore.	*h.s.*	Submerged hyphæ.
c.h.	Carpogonial hypha.	*h.st.*	Sterile hyphæ of sclerotium.
cl.	Columella.	*mc.*	Mucilage.
c.n.	Nutritive branch of carpogonial hypha.	*my.*	Mycelium.
cp.	Carpogonium.	*pl.*	Pollinodium.
cs.	Chlamydospore.	*sg.*	Sterigma.
c.s.	Ascus forming branch of carpogonial hypha.	*sp.*	Sporangium.
en.	Endogonidia (ascospores).	*st.*	Stylogonidia.
ex.	Exosporium.	*zy.*	Zygospore.

PLATE XX.

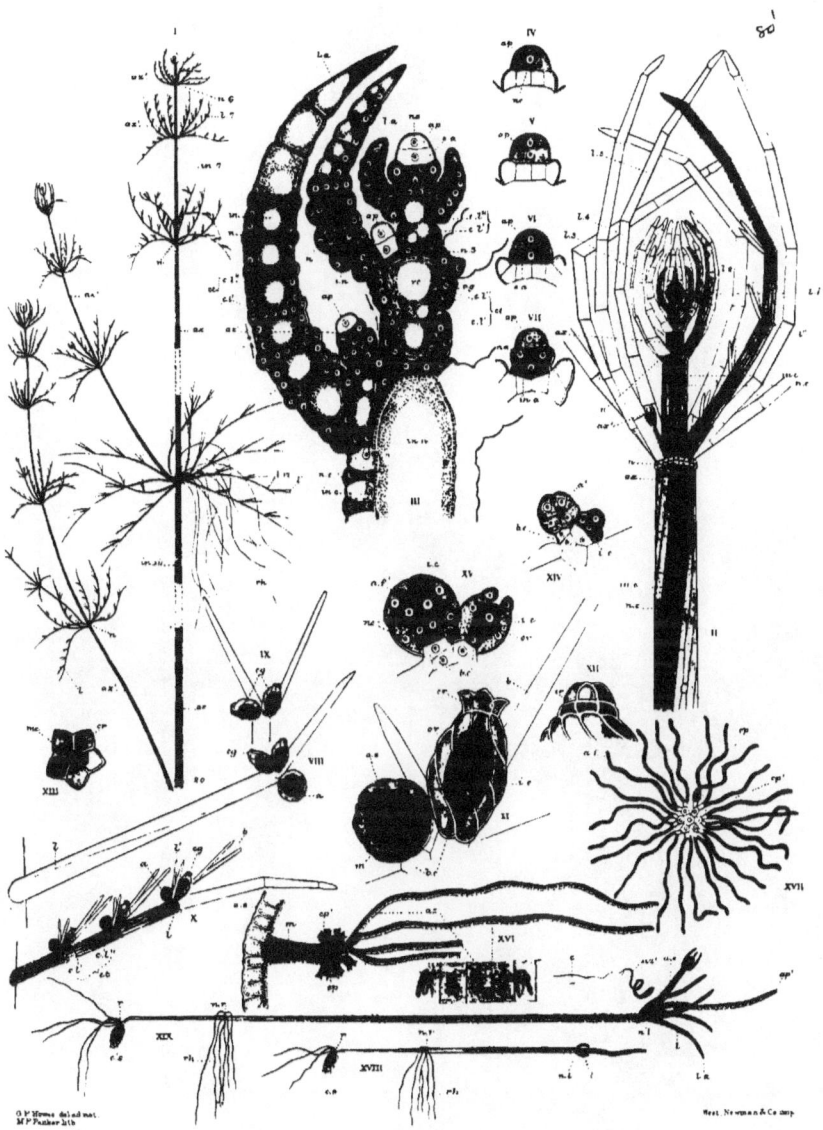

PLATE XX.
THE STONEWORTS.

Fig. I.—An entire actively-growing plant of Chara.*
Growth at the summit was making good disintegration and death at the base, resulting in a separation between the main and the oldest lateral axes, as figured.
The rhizoids are indicated for but one node. Nat. size.

Fig. II.—The same, from internode 5—6 upwards.
The appendages of whorls 2 to 4 were slightly displaced, to render visible the apex. × 8.
One segment of the descending cortical lobe has, in the case of two internodes and of two lateral axes, been shaded darkly. The parts lettered *n.c.* represent one cortical node.
Only one lateral bud is drawn, *ax'*., and no note is taken of the fruits.

Fig. III.—Median longitudinal section through the terminal bud of a similar plant.
The growing apices are left white, and the nodal and cortical elements are shaded darkly. D. 3. Osmic acid and alcohol. See Parker (137).
The arrows indicate the directions of growth of the cortex.
The nuclei of the old cells are not drawn, being in the condition described by Strasburger (142) as fragmentary. See also Johow (135).

Figs. IV. to VII.—Four teased preparations of dividing apical cells, preserved as for Fig. III.
As drawn, they represent in order four successive phases in the developmental history of the growing apex.
The sub-apical (segmental) cell is, in Fig. VI., and its products are in Fig. VII., shaded darkly. D. 3.

Fig. VIII.—A fertile leaf of Nitella, bearing both male and female organs.
Seen from the side. A. 3.

Fig. IX.—A similar leaf, bearing female organs alone. Seen from above. A. 3.

* A synopsis of British Characeæ will be found in Groves (184).

Fig. X.—A fertile leaf of Chara, viewed from the same aspect and drawn to the same scale as that of Nitella, Fig. VIII.

Fig. XI.—One set of reproductive organs from the same, magnified. A. 3.

Fig. XII.—The apex of the carpogonium of Nitella.

Fig. XIII.—The same in Chara, looking down on the micropyle.

Figs. XIV. and XV.—Two stages in the development of the reproductive organs of Chara, in section. From nature after Sachs.
(Figs. VIII., X., XI., XIV., and XV., are all drawn in the same position relative to the axis.)

Fig. XVI.—One antheridial shield of Chara, with its male reproductive apparatus. An immature example, crushed. D. 2.
A few segments of a ripe antheridial filament are drawn below, and to the right one antherozooid, liberated by pressure. F. 3.

Fig. XVII.—The above antheridial shield, etc., seen from within. All the filaments are represented. A. 3.

Figs. XVIII. and XIX.—Two stages in the growth of the pro-embryo; the older one shows the origin of the young plant arising from it,—set free by disintegration of the pro-embryo, as the lateral axes are liberated by that of the adult plant. (Compare Fig. I.)

[For further details and a full bibliography, see Sach's "Text Book of Botany," and Luerssen (136).]

— — —

a.	Antheridium.	c.	Cilia.
a'.	Mother cell of the same.	c.g.	Chlorophyll granules.
a.e.	Axis of young plant.	cy.	Carpogonium.
a.f.	Antheridial filaments.	c.l'.	Ascending cortical lobe.
a.f'.	Parent cells of the same.	c.l''.	Descending cortical lobe.
ap.	Apical cell of axis.	cp.	Capitulum.
ap'.	Apical cell of pro-embryo.	cp'.	Secondary capitula.
a.s.	Antheridial shield.	cr.	Crown of carpogonium.
ax.	Main axis.	c.s.	Carpospore.
ax'.	Lateral axis.	ct.	Cortex.
az.	Antheridial filament, cells of.	i.c.	Investing cells.
az'.	Mature antherozooid.	in.	Internode.
b.	Bracteole.	in.a.	Last formed internode.
b.c.	Basal cell of reproductive organ	in.c.	Cortical internode.

l.	Leaf.	*n.h.*	Leaf-bearing node of pro-embryo.
l'.	Leaflets.	*n.r.*	Root-bearing node of the same.
l".	Sterile leaflets.	*or.*	Oosphere.
l.a.	Apical cell of leaf.	*r.*	Primary root.
m.	Manubrium.	*rh.*	Rootlets (rhizoids).
mc.	Micropyle.	*s.a.*	Segmental cell.
n.	Node.	*rc.*	Sap vacuole.
n.a.	Last formed node.	1 to 5.	Nodes or their appendages, 1 to 5.
n.c.	Cortical node.	*i.* to *v.*	Internodes, i. to v.
nc.	Nucleus.		

THE FERN.

PLATES XXI., XXII.

PLATES XXI., XXII.
THE FERN.

a.	Archegonium.	pc'.	Conjunctive parenchyma.
a'.	Neck of the same.	p.p.	Protophloem of Russow.
a.c.	Apical cell.	pt.	Prothallus.
an.	Annulus.	r.	Roots.
ap.	Growing apex of stem.	r'.	Root hairs.
ap'.	Growing apex of leaf.	r".	Primary root hairs.
at.	Antheridium.	rh.	Rhizome.
at'.	Parent cell of the same.	r.p.	Primary root.
az.	Parent cells of antherozooids.	scl.	Sclerenchyma.
b.c.	Basal cell.	scl'.	Peripheral sclerenchyma.
c.	Cilia (for Fig. XIII. Pl. XXII. Cushion).	scl".	Central sclerenchyma.
ca.	Ventral canal cell.	sg.	Ripe sporangium.
ca'.	Central cell of neck.	sg'.	Young sporangium.
c.c.	Central cell.	sh.	Bundle sheath.
cl.	Chlorophyll grains.	sh'.	Phloem sheath.
ct.	Cotyledon.	sp.	Spores.
f.	So-called foot.	sp'.	Spore mother cells.
i.	Indusium.	s.p.	Sieve plate.
i.c.	Investing cells of sporangium.	st.	Stoma.
i.s.	Intercellular space.	s.t.	Sieve tube.
l.	Leaf stalk (rachis).	tc.	Trichomes.
l'.	Leaflet.	tp.	Tapetum.
l.b.	Lateral bud.	ut.	Protoplasmic utricle.
l.l.	Lateral line.	v.b.	Vascular bundle.
ms.	Mesophyll.	v.b'.	Peripheral bundles.
nc.	Nucleus.	v.b".	Central bundles.
n.c.	Dividing neck cell.	vc.	Vacuole (cell sap).
or.	Oosphere.	xy'.	Protoxylem vessels.
p.	Pinna.	xy".	Scalariform vessels (secondary xylem).
pc.	Parenchyma.		

PLATE XXI.

PLATE XXI.

THE BRACKEN FERN.—External Characters. The Anatomy and Histology of the Vegetative Organs.

Fig. I.—A fern plant, carefully washed after removal from the earth.
The transverse dotted-line indicates the level of the soil, the leaf of the current season ii. alone rises above it, and of those parts beneath it i. represents the leaf next to reach the surface, and iii. the disintegrated base of that of last season. ½ nat. size.
The arrows indicate directions of active growth.

Fig. II.—One node of a fern plant, from which a second leaf has been developed in one season.
i. The past, ii. The current, iii. The young frond. ½ nat. size.

Fig. III.—Transverse section across an internode of the rhizome at its thickest part. The roots are not indicated.

Fig. IV.—Longitudinal section through an internode, at right angles to the lateral line. It passes through the base of a lateral root.

Fig. V.—A transverse section across the youngest node, passing through the axis of a young leaf.
The smaller sclerenchyma tracts are not indicated.
(Figs. III. to V. all × 2.)

Fig. VI.—A portion of the rhizome, from which, after two days' immersion in alcohol, the superficies tissues were scraped away to the level of the bundle system. Nat. size.

Fig. VII.—Portion of a thin transverse section across the rhizome, sufficient being indicated to take in one entire vascular bundle. A. 2.
The bundle sheath, the protophlœm of Russow (138), and conjunctive parenchyma are all shaded over.

Fig. VIII.—Portion of a transverse section of a vascular bundle, more highly magnified. D. 4.

Fig. IX.—Portion of a radial longitudinal section of the rhizome, the same parts being shaded as in Fig. VII. A. 2.

Fig. X.—Portion of a thin radial longitudinal section of a vascular bundle, passing through the protoxylem. This, for the sake of distinction, has been shaded over.
All the vessels are here drawn as seen in surface view. D. 4.

ATLAS OF BIOLOGY.

No attempt has been made in any of the above figures to represent the starch present. It will be found to vary in quantity with the season of the year.

Staining is unnecessary, but if resorted to, hæmatoxylin is preferable.

Fig. XI.—A typical cell of the parenchyma.

Fig. XII.—A cell from the sub-epidermal tissue.
Both of the above were teased up in eosin, from spirit material.

Fig. XIII.—A typical cell of the sclerenchyma.

Fig. XIV.—Portions of a protoxylem vessel.
Torn fragments of the thin, unlignified part of the cell wall are often visible, as at *.

Fig. XV.—A small scalariform tracheid, one portion of which is drawn in section.

Fig. XVI.—A small portion of the boundary wall between two scalariform tracheides, highly magnified. Gundlach's $\frac{1}{15}$th immersion.

Fig. XVII.—A sieve-tube.
(Figs. XI. to XV., and Fig. XVII., are all drawn to the same scale, and in XIV., XV., and XVII., the thickened parts of the cell walls are alone shaded. D. 4.

Figs. XIII., XIV., XV. and XVII. are drawn from cells isolated away under treatment with nitric acid).

Fig. XVIII.—Portion of a sieve-tube from the fresh stem, treated with $\frac{1}{2}$ p.c. osmic acid.

The protoplasmic connections figured are inserted from De Bary (132). Gundlach's $\frac{1}{15}$th immersion.

Fig. XIX.—Median longitudinal section through the apex of the rhizome.
The apical mass of indifferent tissue which graduates off into the several constituents of the stem, is shaded darkly. A. 2.

Fig. XX.—A portion of the above highly magnified, showing the apical cell, and the two cells of the indifferent tissue last cut off from it. Of these, the older had subsequently divided up. D. 4.

Fig. XXI.—Transverse section across a similar apex. Details as for Fig. XX. D. 4.

PLATE XXII.

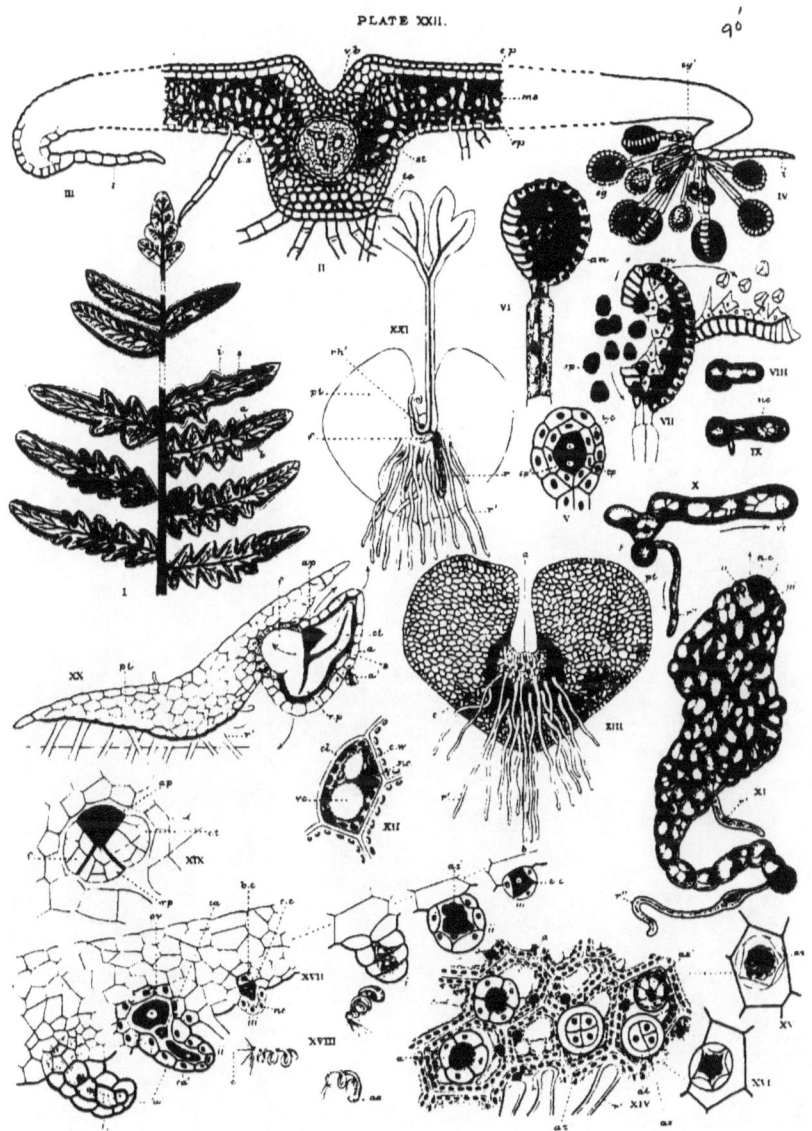

PLATE XXII.

THE BRACKEN FERN.—THE LEAF, REPRODUCTIVE ORGANS, AND DEVELOPMENT.

FIG. I.—Portions of the same leaf, seen from beneath. × 2.

FIG. II.—Section across the plane *a-b*. of one pinnule of the above. D. 2.
The chlorophyll-bearing cells are alone shaded.

FIG. III.—Similar section across the margin of a sterile leaf. D. 2.

FIG. IV.—A similar section across the margin of a spore-bearing leaf. D. 2.

FIG. V.—A young sporangium from the above, in optical section.

FIG. VI.—A ripe sporangium from the same. Surface view.

FIG. VII.—A similar sporangium after rupture under pressure.
The right hand portion of the figure, indicated in outline, represents the result of rupture under imbibition of water.
(Figs. V. to VII. D. 3.)

FIGS. VIII., IX., X.—Early stages in the germination of a fern-spore, grown on a slab of limestone in a warm damp atmosphere. (The outside of a flower pot answers very well). D. 4.
No trace of chlorophyll was yet visible.

FIG. XI.—A later stage of the same.
The developing prothallus, now quite green, was growing from an apical cell, *ap*.
i. to iii. indicate the last formed cells of the apex. (Compare Figs. XX. and XXI. of Plate XXI.) D. 4. ½ p.c. osmic acid.

FIG. XII.—One cell of Fig. XI. in detail, viewed in optical section. F. 2.

FIG. XIII.—A sexually mature prothallus, viewed from beneath.
The archegonia are indicated in black, and the antheridia are left white. × 10.

FIG. XIV.—A portion of the same near the edge of the cushion, highly magnified.
The antheridia are drawn focussed to the level of the dividing central cells, and the detailed structure of the investing cells of both archegonia and antheridia is omitted.

FIG. XV.—An older antheridium.

Fig. XVI.—Another, after the discharge of the antherozooids.
(Figs. XV. and XVI. were near the margin of the prothallus. Figs. XIV. to XVI. D. 4.)

Fig. XVII.—Section across the sexually mature prothallus, through the plane a-b, of Fig. XIII.

The left half* of the figure represents three views of the unfertilized archegonium in *Aspidium*, and the right half corresponding aspects of the antheridium in *Pteris*. In each case they are—

    i. The ripe organ, surface view.
    ii. The same in section.
    iii. The young organ, in section. D. 3.

No attempt has been made to fill in the detailed structure of individual cells. The archegonia not unfrequently occur on the upper side of the prothallus as well as on the under.

Fig. XVIII.—Three ripe antherozooids, after liberation under pressure.
$\frac{1}{2}$ p.c. osmic acid. Gundlach's $\frac{1}{16}$th immersion.

Fig. XIX.—The segmenting egg-cell of *Asplenium*.
After Schenk (139), vol. i., p. 217.
The first formed septa are indicated by a thick black line.

Fig. XX.—Section across an embryo-bearing prothallus of *Nephrolepis*, cut as directed for Fig. XVII.* A. 3.

In Figs. XIX. and XX. homologous parts are shaded alike, and the arrows indicate the directions of growth of the four subdivisions of the embryo.

    * Marks the line of differentiation of the vascular bundle.

For *Pteris*, similar stages to the above have been figured by Hofmeister and others. They are reproduced in Luerssen (136), vol. i.

See also Hofmeister's papers on the higher *Cryptogams*. Translated by F. Currey. *Ray Soc. Publications*, 1862.

Fig. XXI.—A fern-bearing prothallus, seen from beneath. × 8.
The central dark line indicates the bundle system, and the parts of the embryo are shaded as for Figs. XIX. and XX.

    * For the sections from which these were drawn I am indebted to my friend Mr. F. O. Bower.

THE FLOWERING PLANT.

PLATES XXIII., XXIV.

PLATES XXIII., XXIV.
THE FLOWERING PLANT.

a.,	Apex of main axis.	nc.	Nucleus.
a'.	Apex of root.	nu.	Nucellus.
a.b.	Axillary bud.	o.	Ovule.
a.c.	Air cavity.	oe.	Oosphere.
an.	Annular vessels.	oy.	Ovary.
ap.	Antipodal cells.	p.	Mouths of pits.
ax.	Main plumular axis.	pc.	Parenchyma of cortex.
ax'.	Hypocotyledonary stem.	pn'.	Parenchyma of pith.
ax".	Main or primary root.	ph.	Phloem (Protophloem, in Figs. XXIV. and XXV., Pl. XXIII.)
az.	Pollen grains.		
az'.	Mother cells of the same.	ph'.	Hard bast.
c.	Cuticle.	ph.r.	Phloem rays.
cb.	Procambium.	pl.	Pallisade tissue.
cb'.	Cambium.	p.p.	Phloem parenchyma.
cb".	Pericambium.	pt.	Petals.
cp.	Carpels.	p.t.	Pollen tube.
ct.	Cotyledon.	p.r.	Pitted vessels.
ct'.	Base of attachment of the same.	r.	Secondary roots.
c.w.	Cell wall.	r".	Lateral rootlets.
cx.	Cortex.	s.	Stoma.
e.	Embryo.	sb.	Subsidiary cells.
ed.	Bundle sheath (endodermis).	se.	Sepals.
en.	Endosperm.	sg.	Stigma.
en'.	Parent endosperm nucleus (central nucleus of embryo sac).	s.l.	Spiral layer of anther.
		sm.	Stamens.
ep.	Epidermis.	sp.	Spiral vessels.
ep'.	Subepidermis (collenchyma).	s.p.	Sieve plate.
e.s.	Embryo sac.	st.	Style.
f.	Funiculus.	s.t.	Sieve tube.
g.c.	Guard cells.	su.	Suspensor.
gl.	Glandular hairs.	su'.	Basal cell of suspensor.
gr.	Starch grains.	sy.	Synergidae.
gr'.	Aleurone grains.	t.	Testa.
i.	Integuments.	tc.	Trichomes.
i.c.	Intercellular space.	tp.	Tapetum.
i.n.	Internode.	ut.	Primordial utricle.
l.	Leaf.	v.	Vacuole.
mc.	Micropyle.	v.b.	Vascular bundle.
md.	Medulla (pith).	x.p.	Xylem parenchyma.
md'.	Pith cavity.	xy.	Xylem.
m.r.	Medullary rays.	xy'.	Protoxylem.
ms.	Mesophyll.	xy".	Xylem sclerenchyma.
n.	Node.	xy.r.	Xylem rays.

PLATE XXIII.

PLATE XXIII.

THE FLOWERING PLANT.—External Characters. The Vegetative Axis.

Unless otherwise stated, all the figures on Plates XXIII. and XXIV. illustrate the structure of the French Bean plant, *Phaseolus vulgaris*.

Fig. I.—The entire seed, viewed from its attached surface, after a few hours' immersion in water.

Fig. II.—The same, after removal of the testa.

Fig. III.—The same, after removal of the right cotyledon.
Seen from within.

Fig. IV.—A similar seed during germination, just on leaving the soil.
Grown in damp sawdust, the surface level of which is indicated by the line i—ii.
(Figs. I. to IV. all nat. size.)

Fig. V.—A twelve-days' plant, grown as for Fig. IV.
The unlettered transverse line represents the surface level of the sawdust.
The arrows indicate the direction of growth of the original main axis.
The cotyledons, colourless in Fig. IV., were here quite green. $\frac{1}{2}$ nat. size.

Fig. VI.—Median longitudinal section through the embryonic axis of Fig. III.
The arrows indicate the direction in development of the procambium of one lateral axis, differentiation having commenced at the point marked by the adjacent line.
Picric acid and alcohol. Eosin stained, preserved in Canada balsam. A. 2.

Fig. VII.—A portion of the hypocotyledonary axis of the above. D. 3.
The several tissues figured, graduate into each other as the apex is reached.

Fig. VIII.—Section of a cotyledon of the above. F. 2. Fresh, stained eosin.
It is important to guard against mistaking cut corners of cells for intercellular spaces.

Fig. IX.—The apical cone of Fig. V. teased out with needles.
The internodes are shaded darkly. D. 3.
The apex of *Anacharis* may profitably be teased up for comparison.

Fig. X.—Transverse section across the main root of Fig. V. Nat. size.

Fig. XI.—Median longitudinal section of the epicotyledonary axis of Fig. V.   × 10.

Fig. XII.—Transverse section of the same, across the point marked *—* in Fig. V. × 10.

Fig. XIII.—A similar section across the same axis of an older plant.   × 5.
(In Figs. XI. to XIII. the cambium layer is represented as a black line.)

Fig. XIV.—The bundle-system of the French Bean, seen from the side. Dissected as for the Fern, Fig. VI., Plate XXI. From nature, after Nageli. See De Bary (132).*
The bundles of the left side are shaded darkly.   × 2.

Fig. XV.—Thin transverse section of a portion of Fig. XII.
The protoxylem elements, xy'., may often be at once recognized in transverse sections, as the spiral or annular thickenings frequently tear away in cutting and hang ragged. D. 3.

Fig. XVI.—Radial-longitudinal section of a similar stem, taken through the axis of a young bundle. D. 3.

Fig. XVII.—Portion of a thin transverse section of Fig. XIII.
The phloem parenchyma and the rays of the vascular bundle are here shaded. D. 2.

In Figs. XV. and XVII., the trichomes are for the most part omitted, and no note is taken of the chlorophyll present in the cells of the parenchymae.
The only structures in which the protoplasm is drawn, are the sieve tubes; it is represented in them in the contracted post-mortem state. (See below.)

Fig. XVIII.—Sections of the sieve tube in the above.
i.   The protoplasm, drawn as in life.
ii. The protoplasm, drawn in the shrunken post-mortem condition.
iii. Portion of a sieve-tube bearing both transverse and lateral sieve plates. Iodine. Gundlach's $\frac{1}{15}$th immersion.

Fig. XIX.—A cambium cell, isolated under treatment with nitric acid, as for the fern-stem, Figs. XIV., etc., Plate XXI. D. 3.

Fig. XX.—A similar cell, as seen in tangential section of the cambium layer. D. 3.

Fig. XXI.—A dividing cambium cell, drawn from the layer, cb'., of Fig. XVI., as seen in radial-longitudinal section. D. 3.

* Large models of this and certain other bundle-systems, after Nageli and others, are made by H. Gassor, of the Botanical Institute, Graz.

THE FLOWERING PLANT.

FIG. XXII.—A portion of an isolated pitted vessel, seen *en face*.

FIG. XXIII.—Section through the partition wall between two pitted vessels. Gundlach's $\frac{1}{15}$th immersion.

FIG. XXIV.—Transverse section across i.—ii. of the main root of Fig. V., after hardening in alcohol.
The section passed through a young lateral root. A. 2.

FIG. XXV.—The greater portion of the same, more highly magnified.*
No note is taken of the intercellular spaces. D. 3.

FIG. XXV.ª—The bundle-sheath of the above enlarged. Gundlach's $\frac{1}{15}$th immersion.

* Sections such as this, to show the typical arrangement of the parts of the root, must be cut low down near the apex.

PLATE XXIV

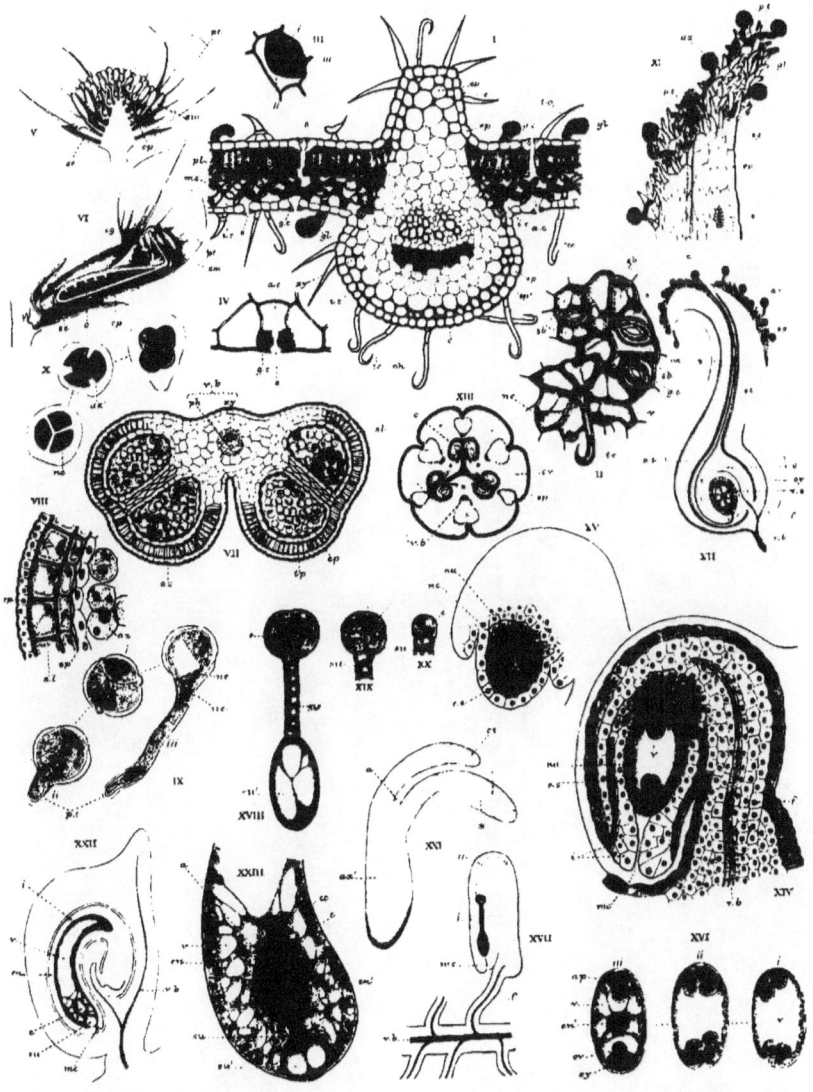

PLATE XXIV.

THE FLOWERING PLANT.—THE LEAF AND FLOWER. THE REPRODUCTIVE ORGANS, AND DEVELOPMENT.

FIG. I.—Transverse section across the oldest leaf of a young bean-plant. The whole leaf was rolled up and cut in the fresh state. D. 3.

FIG. II.—A portion of the epidermis of the same stripped off, showing the characters and relations of the stoma and guard-cells, the subsidiary cells, and the trichome. D. 4.
The glandular hairs arise as central cells of a series, similarly to the trichome figured.
Examined fresh in water, the details of cell-structure being drawn after treatment with 1 p.c. osmic acid.

FIG. III.—Slightly diagrammatic representation of a developing stoma.
The entire parent cell is drawn, and the septa by which it was divided up are numbered i. to iii., in order of development; iii. would have finally split to form the stoma.
The specimen figured was obtained from near the edge of a young leaf. Fresh. F. 3.

FIG. IV.—Section through a stoma of Fig. I. F. 4.

FIG. V.—Median longitudinal section of a *Buttercup*. × 3.

FIG. VI.—Similar section of a Bean-flower. × 2.

FIG. VII.—Transverse section of the anther of a nearly matured *Buttercup*. Alcohol and weak glycerine. D. 2.
By cutting the whole flower-head between the finger and thumb, most instructive sections may be obtained, cut at all possible levels.

FIG. VIII.—A small portion of Fig. VII., more highly magnified. F. 3.

FIG. IX.—Three pollen-grains, teased out from the stigma of an open *Buttercup*. (See Fig. XI.)
Taken in the order numbered, they represent stages in the development of the pollen tube. Such stages may be obtained with ease, by object-glass culture in a saccharine fluid. 1 p.c. osmic acid. D. 3.
The smaller nucleus is stated by Strasburger (143) to be alone the active agent in fertilization,

FIG. X.—Developing pollen-grains, from the anther of a young *Buttercup*. Teased. F. 3.

FIG. XI.—Surface view of a thick longitudinal section, through the stigma of an open *Buttercup*.
The central shaded portion * represents the central cavity of the style. Alcohol and weak glycerine. D. 3.

FIG. XII.—Slightly diagrammatic longitudinal section of the entire carpel of an open *Buttercup*, at the period of fertilization.
This figure represents the facts, as pieced together from observations made upon a number of sections, obtained as for Fig. VII.
* Indicates a single well-differentiated layer of cells, surrounding the central canal of the style.

FIG. XIII.—Transverse section of the carpel of an unopened flower of *Lilium*, to show the general relations of the ovules, etc.
The parts indicated in black represent the passage open to the pollen tube, when once it has entered the central canal of the style. A. 2.

FIG. XIV.—An ovule from the same, more highly magnified. D. 2.

FIG. XV.—A similar section from the carpel of a much younger flower. D. 3.

FIG. XVI.—Three embryo sacs from the same carpel as Fig. XIII., to show the changes undergone prior to fertilization. D. 2.
As numbered in order i. to iii. they represent successive stages in the development of an individual. Figs. XIV. and XV. present still earlier phases in the same development.
The above sections (Figs. XIII. to XVI.) were prepared from fresh material, transferred as cut into a half-and-half solution of methylated spirit and glycerine, and kept exposed in a warm room until the spirit had evaporated.

FIG. XVII.—An ovule, teased up from the fruit of the *Shepherd's Purse*. A. 3.

FIG. XVIII.—The embryo drawn in Fig. XVII. i. liberated under gentle pressure. D. 4.

FIGS. XIX. and XX.—Two embryos, teased from younger fruits of the same plant. D. 4.
The three embryos above figured represent successive stages in development, and only such are here drawn as can be readily seen with ease by teasing up the fruits with needles.
For further details see Hanstein's figures, reproduced in all the text-books.

FIG. XXI.—A much older embryo of the same plant, at this stage quite green.
The curvature of the cotyledons becomes intelligible on considering the natural position of the embryo, shown at ii. in Fig. XVII. In life, the cotyledon marked* overlay its fellow, as indicated by the dotted lines in Fig. XVII., thus hiding the apex from view. A. 3.

FIG. XXII.—Median longitudinal section of the fruit of *Potamogeton* (the Broad Pondweed), to show the relations of the endosperm to the embryo. Alcohol and weak glycerine. A. 2.

Each fruit of the above plant contains but one embryo, and if, holding the fruits between the finger and thumb, the sides are sliced away, most satisfactory preparations can be readily obtained.

FIG. XXIII.—A portion of the above more highly magnified, showing the nuclei of the endosperm, and the embryo in detail. D. 4.

The distinctive characters of the *Monocotyledon* embryo are here drawn, as compared with those of the *Dicotyledon* represented in Fig. XXI.

APPENDIX
AND
BIBLIOGRAPHY.

APPENDIX.

A. The terms right and left are always used in anatomy, with reference to the right and left sides of the subject's body. Similarly, by the anterior is always meant the head end, and by the posterior the hinder end. The terms dorsal and ventral apply to the back and the belly respectively; they are, however, often replaced by terms bearing directly upon the position of the nervous axis, viz., neural and hæmal for the vertebrate, hæmal and neural for the invertebrate.

B. The method of dissection from the side here employed has long been recognized as yielding results second to no other, where a general survey of the animal's entire anatomy is required. Moreover, by cutting thus to one side of the middle line, the attachments of the organs are not interfered with, and they, therefore, remain in position for detailed examination.

In dissecting all the animals dealt with in this work, some such stand as that figured above will be found useful; the rests for the arms may conveniently be lengthened to suit individual requirements, but the swinging-arm, carrying a watchmaker's lens, will be found a necessity in examining delicate structures.

All dissections should be performed under water, with the subject in a perfectly rigid position. For this purpose a dish, such as is here figured, is necessary; it should be preferably of glass, one-third filled with paraffin blackened by the admixture of a little lamp-black. A slab of the above material admits of the subjects being pinned down, and it should always be weighted at the bottom with lead, to avoid floating up.

Should the water become clouded in dissecting, it must be at once changed, the cause of the obstruction being, if possible, washed away.

While dissecting, it is highly desirable that living specimens of the animal under consideration should be kept constantly under observation, as many a structural feature is only intelligible on a knowledge of the creature's habits.

C. Whole animals, or dissections, are best preserved in alcohol. The ordinary methylated spirit of commerce, diluted with ⅓rd its bulk of water, answers admirably; delicate structures, such as the brain, must be placed at once into strong methylated spirit, and in all cases the preservative medium should be replaced after the first twenty-four hours. With the above exceptions, it is not desirable to submit any preparation at once to the action of strong methylated spirit, as such a course is apt to result in a too rapid dehydration of the superficial parts, thus preventing that complete permeation necessary for successful preservation.

Where very careful preservation without shrinking is desired, it is well to use in succession, 50 p.c. for twenty-four hours, 75 p.c. for the next forty-eight, methylated spirit *ad infinitum*, being careful to prevent the access of air.

D. For ordinary coarse injection of the blood vessels, a mixture of French blue, in the proportions of a teaspoonful to a tumbler of water, cannot be surpassed. It is cheap, sufficiently finely divided to enter all but the capillary vessels in a small animal, and it can be used with little trouble. Moreover, the fact that it is insoluble in water renders it the more valuable, as injection can be carried on piecemeal as circumstances may require.

In dealing with the animals adopted for this book, a small cannula, provided with a tight-fitting india-rubber ball or nipple, as figured below, is all-sufficient as a syringe.

For histological purposes, a gelatine injection is best suited, made as follows :—Allow a given quantity of Nelson's gelatine to stand for twenty-four hours in twice its bulk of water, boil and stir well, adding either French blue or vermilion to colour as required. This fluid should be injected at a temperature such as the hand can comfortably bear.

In preparing a Frog for injection with gelatine, an incision should first be made along the whole ventral integument (care being taken to injure nothing else), and the two halves reflected. Raising the xiphoid process with a pair of forceps, next insert the point of the scissors under this, and carefully remove the whole ventral portion of the shoulder-girdle. By this means the heart will be exposed without injury, the pericardium should now be laid open, and an incision made into the apex of the ventricle, to allow the escape of as much blood as possible. When the animal is sufficiently well bled, a canula* should be tied into the heart, and connected, by means of an inch or two of india-rubber tubing, with the syringe, injection being performed under a gentle, steady pressure.

When the syringe is removed, the tube by which it was connected to the canula should be plugged with a piece of glass rod, and the whole animal placed in water for two or three hours, that the injection may set. After the above treatment, the parts to be preserved should be dealt with as recommended for the uninjected tissues.

E. The term branch as applied to blood-vessels is often very vaguely employed. Defining an artery as a vessel carrying blood from the heart, and a vein as one conveying it to the heart, it will be well to restrict the term *branch* to those vessels formed by the breaking up of an artery into smaller trunks, and that of *factor* to those uniting to form larger veins; arteries and branches being efferent, factors and veins being afferent, as related to the heart.

F. By a wet preparation is meant one that has never been allowed to dry.

The skeleton of the Frog may be readily prepared by allowing the body to lie in water for a few days, after having first removed the skin, viscera, and as much of the flesh as possible. Maceration having gone as far as is desirable, the whole should be put under running water for a day, before it is allowed to dry.

* This must be made to the required size by drawing out a piece of glass tubing, either in a flame or by means of the blowpipe. It should be slightly constricted near the apex, to allow of its being firmly tied into the heart.

APPENDIX.

The process of maceration may be materially hastened by using warm water instead of cold; this method should, however, be adopted with caution, as the parts rarely hold together so successfully as under the first-named treatment.*

When careful preparation of both bony and cartilaginous parts is needed, as in the case of the skull figured in this work, the only reliable method is that of dissecting away the soft parts under water, preserving the whole in spirit as in the case of an ordinary dissection.

G. By far the most successful preservative fluids for histological work are alcohol and picric acid; of the latter, either a cold saturated solution or Kleinenberg's preparation will suffice for the purposes of this work.†

The tissue to be preserved should be removed from the body as soon as possible after death, cut up into small pieces, and put at once into the preservative medium. For most tissues six to eight hours' immersion will be found to suffice, after which the preparation must be transferred to alcohol of gradually increasing strength, viz., 50 p.c., 75 p.c., methylated spirit, and absolute alcohol. Should any excess of acid be present, as can readily be seen from the colour of the spirit, the latter must be repeatedly renewed until it is removed. When this stage is reached the tissue is ready for cutting, and if preserved longer should be kept in absolute alcohol.

Where the tissues are transferred at once from the body to alcohol, the method of treatment should also be as above stated; if osmic acid is used, the preparation should be transferred from it to 50 p.c. alcohol, as soon as it begins to assume a black tint. The requisite length of exposure to osmic acid will be found to vary for different tissues, and experience alone can enable the student to use this reagent with success. Where decalcification is necessary, the tissue to be operated upon should be placed for twenty-four hours in $\frac{1}{4}$ p.c. solution of chromic acid, then for a similar time in 1 p.c. solution, this being either renewed or replaced by a still stronger when necessary. Decalcification completed, the tissue must be transferred to alcohol as above.

For purposes of staining there is no reagent which gives such uniformly good results as Grenacher's solution of borax carmine. The preparation may be allowed to remain in this for a couple of days, without fear of overstaining. When stained, it should be again transferred to methylated spirit and absolute alcohol, before imbedding, in order that any water or excess of staining fluid may be got rid of.

Imbedding.—The best imbedding material is paraffin, preferably that which shall melt at from $50°$ to $60°$ C.

A block of this substance of the calibre of a candle, and about an inch and a half in length, will suffice for all requirements here needed. A pit should be made at one end, large enough to take the preparation with ease. The preparation, previously soaked in turpentine to saturation, should be first transferred to paraffin, the temperature of which must not exceed that of its melting-point, and allowed to remain in the same until permeated thereby; it must next be finally transferred to the block prepared to receive it, and completely covered in paraffin, the whole being then allowed to cool.

Section cutting.—For the purposes of this book an ordinary razor will suffice with which to cut sections; but where a successional series of slices are required, recourse must be had to one of the many microtomes now in use.

Before cutting the sections, it is desirable to remove as much of the imbedding material as possible from around the preparation.

When once mastered, the following is the least laborious of all methods of cutting and mounting.

---

* The results of a long experience, in this and other matters which pertain to the preparateur's art, will be found embodied in Wilder and Gage's "Anatomical Technology," New York and Chicago, 1882.

† The mode of preparation of these and other preservative and staining reagents, will be found fully stated in Huxley and Martin and other similar works, and the student will find of great service a list of those reagents which have yielded such admirable results in connection with the zoological station, Naples. See *Jour. R. M. Soc.*, Series 2, vol. ii., 1882.

ATLAS OF BIOLOGY.

Transfer the sections as cut, together with the imbedding material in which they lie, to a slide, the surface of which has been previously painted over with a heated solution of white shellac in kreasote. Submit the whole to the temperature of the melting-point of the paraffin, until the kreasote is evaporated, when the sections will become firmly adherent to the shellac. The slide must next be immersed in turpentine, which will dissolve up the remaining imbedding material, thus leaving the sections fixed in place and ready to be covered with Canada balsam.

For final mounting well-powdered Canada balsam, first thoroughly dried, must be dissolved in benzole or chloroform, to the consistency of glycerine. A drop of the balsam thus prepared should be deposited upon the slide by means of a glass rod, and allowed to diffuse itself among the sections; the cover-slip should then be let down gently and obliquely upon the objects, its under surface first having been smeared with Canada balsam.

Under the above method of preparation the sections may be cut with a dry razor, but where, as in the case of vegetable stems, etc., they are to be taken from material which has not been submitted to any such treatment, both razor and preparation must be kept moistened during the process with the fluid in which the preparation has been preserved.

Fresh preparations of animal tissues, when not frozen, must always be examined in serum or normal salt solution. For the examination of fresh vegetable tissues, water will suffice.

The microscope used in making the drawings for this volume was one by Zeiss of Jena, and in the formulæ given in the text (example D. 2) the letter refers to the eye-piece, and the numeral to the objective employed.

In no case should a high-power eye-piece be resorted to unless absolutely indispensable, as the sharpness of definition obtained by low ones, such as Zeiss 2 or 3, is highly desirable.

A micrometer of some kind is indispensable to the student of histology, and the purpose to be aimed at in its use, is that of a knowledge of the size of the objects under examination, rather than that of merely ascertaining the magnifying power of the microscope. Most satisfactory results are to be obtained by using an eye-piece micrometer ruled in squares, and by drawing both these and the object to the proportions observed. Having thus a record of objects, drawn in proportion relative to squares of unknown value, it remains but to ascertain the dimensions of these. This is best done by using, in conjunction with the eye-piece micrometer mentioned above, a stage micrometer, ruled to known intervals, whereupon there will be superposed lines of known and unknown value. The actual value of the squares of the eye-piece micrometer may now be once for all calculated, and a record kept of the same for each lens combination.

H. Preparations of shells and similar hard parts are best made as follows:—

A small piece of the structure to be prepared should be first isolated, and then cemented with Canada balsam to a piece of plate glass. When quite set it can be ground down on a rough surface to the required thinness, and finally dislodged for mounting, by submerging the whole in benzole. It may then be put up in Canada balsam in the usual manner.

I. For object-glass culture it is best to use either a small cell of the ordinary type, or a slide which is excavated. Filling this with the nutritive fluid—water for protococcus, Pasteur's fluid for the moulds, sugar-solution for the pollen-grain, etc.—next introduce the organisms upon which observation is to be made, and place a cover-slip over the whole, the edges of which should be oiled to check evaporation. If the observation be a prolonged one, it is well to place the nutritive medium in communication with a reserve of the same fluid, by means of a cotton thread.

APPENDIX.

J. The animals dealt with in this work are best killed as follows :—

The *Frog*.—Place the animal beneath an inverted tumbler, together with a piece of cotton-wool an inch square, saturated with chloroform, and cover the whole with a cloth.

The *Crayfish*.—Easily killed under chloroform as above.

Crayfish may be kept alive and active for some days in a moist atmosphere, and fed upon sopped bread. They thrive much better in captivity under this treatment, than any other known to me.

The *Earthworm*.—Earthworms may be killed, with the body naturally extended, by immersing them for two minutes in methylated spirit, and then in running water for half-an-hour.

The *Snail*.—Snails can be killed with their tentacles often fully extended, by placing them in water as hot as the hand will comfortably bear it.

The *Mussel*.—Place the animal in cold water, and heat slowly to 40° C. The foot will generally be well protruded at death, under this treatment.

BIBLIOGRAPHY.

LIST OF ABBREVIATIONS.

Ann. N. Hist.	Annals and Magazine of Natural History.	London.
Ann. Sci. Nat.	Annales des Sciences Naturelles.	Paris.
Arb. Würz.	Arbeiten aus dem Zoologisch-Zootomischen Institut der Universität.	Würzburg.
Archv. Mk. Anat.	Archiv für Mikroskopische Anatomie.	Bonn.
Archv. Mus. Paris.	Nouvelles Archives. Muséum d'Histoire Naturelle.	Paris.
Archv. Z. Exp.	Archives de Zoologie Expérimentale.	Paris.
Ber. Gies.	Berichte der Oberhessischen Gesellschaft für Natur-und Heilkunde.	Giessen.
Bot. Zeit.*	Botanische Zeitung.	Leipzig.
Cpt. Rend.	Comptes Rendus de l'Académie Française.	Paris.
Denk. Kais. Akad.	Denkschriften der Kaiserlichen Akademie der Wissenschaften.	Vienna.
Enc. Brit.	Encyclopædia Britannica.	9th Edition.
Jour. Anat.	Journal of Anatomy and Physiology.	London and Cambridge.
Jour. R. M. Soc.†	Journal of the Royal Microscopical Society.	London.
Mem. Akad. St. P.	Mémoires de l'Académie Impériale des Sciences.	St. Petersburg.
Morph. Jahrb.	Morphologisches Jahrbuch.	Leipzig.
Phil. Trans.	Philosophical Transactions of the Royal Society.	London.
Pr. R. Soc.	Proceedings of the Royal Society.	London.
Pr. Z. Soc.	Proceedings of the Zoological Society.	London.
Q. J. M. S.	Quarterly Journal of Microscopical Science.	London.
Rev. Sci.	Revue Scientifique.	Paris.
Szb. Bonn.	Sitzungsberichte der Niederrheinischen Gesellschaft für Natur- und Heilkunde.	Bonn.
Szb. Wien.	Sitzungsberichte der Kaiserlichen Akademie der Wissenschaften.	Vienna.
Zeitsch. Jena.	Jenaische Zeitschrift für Medicin und Naturwissenschaft.	Leipzig and Jena.
Zeit. W. Zool.	Zeitschrift für Wissenschaftliche Zoologie.	Leipzig.
Zool. Anz.*	Zoologischer Anzeiger.	

* A systematic record, giving the full titles of all monographs, etc., produced from time to time in its own branch of natural science, will be found in this periodical.

† Brief translations and abstracts of certain of the more important biological papers published, exclusive of those on gross vertebrate anatomy, are issued with this Journal.

FROG.

1. BLOOMFIELD, J. E.—On the Development of the Spermatozoon in Helix and Rana. *Q. J. M. S.*, vol. xxi., 1881.
2. BOAS, J. E. V.—Ueber den Conus Arteriosus, und die Arterienbogen der Amphibien. *Morph. Jahrb.*, vol. vii., 1881.
3. BOURNE, A. G.—On Certain Abnormalities in the Common Frog. *Q. J. M. S.*, vol. xxiv., 1884.
4. DE WATTEVILLE, A.—A Description of the Cerebral and Spinal Nerves of Rana esculenta. *Jour. Anat.* vol. ix., 1875.
5. ECKER, A.—Icones Physiologicæ. *Leipzig*, 1851-9.
6. ECKER, A.—Die Anatomie des Frosches. Part I. *Brunswick*, 1864.
7. FÜRBRINGER, M.—Zur Vergleichenden Anatomie und Entwicklungsgeschichte der Excretionsorgane der Vertebraten. *Morph. Jahrb.*, vol. iv. 1878.
8. GOETTE, A.—Die Entwicklungsgeschichte der Unke. *Leipzig*, 1875.
9. HOFFMANN, C. K.—Bronn's Klassen und Ordnungen des Thierreichs. Vol. vii., Amphibien. *Leipzig und Heidelberg*, 1873-8.

BIBLIOGRAPHY.

10. HOFFMANN, C. K.—Beiträge zur Kenntniss des Beckens der Amphibien und Reptilien. *Leiden*, 1876.
11. HUXLEY, T. H.—Article "Amphibia." *Enc. Brit.*, 9th Edit.
12. JOURDAIN, A.—Recherches sur le système lymphatique de la Rana temporaria. *Rev. Sci.*, 1883. See also papers by Langer, *Stzb. Wien*, 1866. 7, 8, and *Ann. N. Hist.*, 1868.
13. KLEIN AND NOBLE SMITH.—Atlas of Histology, *London*, 1879-80.
14. KNAUER, F. K. — Naturgeschichte der Lurche. *Vienna*, 1878.
15. LEYDIG, F.—Die Anuren Batrachier der Deutschen Fauna. *Bonn*, 1877.
16. MARSHALL, A. M.—Certain Abnormal Conditions of the Reproductive Organs in the Frog. *Jour. Anat.*, vol. xviii., 1884.
17. MÜLLER, W.—Ueber Entwicklung und Bau der Hypophysis und des Processus infundibuli cerebri. *Zeitsch. Jena.*, vol. vi., 1871.
17A. OSBORNE, H. J.—Observations upon the Urodele Amphibian Brain. *Zool. Anz.*, vol. vii., 1884.
18. PARKER, W. K.—On the Structure and Development of the Skull of the Common Frog. *Phil. Trans.* 1871.
19. RANVIER, L.—Technisches Lehrbuch der Histologie, *German Translation. Leipzig*, 1877.
20. RAUBER, A.—Neue Grundlegungen zur Kenntniss der Zelle. *Morph. Jahrb.*, vol. viii., 1882.
21. SEDGWICK, A.—Early Development of the Wolffian Duct., etc. *Q. J. M. S.*, vol. xxi., 1881.
22. SPENGEL, J. W.—Das Urogenitalsystem der Amphibien. *Arb. Würz.*, vol. iii., 1876.
23. WHITNEY, W. N.—The Changes which accompany the Metamorphoses of the Tadpole, etc. *Q. J. M. S.*, vol. vii., 1867.
24. WIEDERSHEIM UND ECKER. — Die Anatomie des Frosches. Parts ii. and iii. *Brunswick*, 1881-2.

CRAYFISH.

25. ALBERT F. V.—Das Kaugerüst der Dekapoden. *Zeit. W. Zool.*, vol. xxxix., 1883.
26. GERSTAECKER, A.—Bronn's Klassen und Ordnungen des Thierreichs. Vol. v., Arthropods. *Leipzig*, 1866 to date.
27. HERRMANN, G.—Spermatogénésis de Crustacés Podophthalmiques. *Cpt. Rend.*, 1881.
28. HUXLEY, T. H.—On the Classification and Distribution of the Crayfishes. *Pr. Z. Soc.*, 1878.
29. HUXLEY, T. H.—The Crayfish, etc. *International Scientific Series*, 1880.
30. JOURDAIN, S. — Sur les cylindres sensoriels de l'autenne interne des Crustacés. *Cpt. Rend.*, 1880.
31. MOCQUARD, M. F. — Recherches anatomiques sur l'estomac des Crustacés podophthalmaires. *Ann. Sci. Nat. Zool.*, 1883.
32. RATHKE, H.—Ueber die Bildung und Entwicklung des Flusskrebses. *Leipzig*, 1829.
33. REICHENBACH, H.—Die Embryonanlage und erste Entwicklung des Flusskrebses. *Zeit. W. Zool.*, 1877. *Abstract by T. J. Parker, Q. J. M. S.*, vol. xviii., 1878.
34. RUTHERFORD.—On the Structure of Arthropod Muscle *Pr. R. Soc. Edinburgh*, 1883.
35. SARS, G. O. — Om Hummerens postembryonale udvikling. *Christiana*, 1874. See also a paper by S. J. Smith, in *Trans. Connecticut Acad. of Arts and Sciences*, vol. ii., 1873.
36. WASSILIEW, E.—Ueber die Niere des Flusskrebses. *Zool. Anz.*, vol. i., 1878.
37. WEBER MAX.—Ueber den Bau und die Thatigkeit der sogenannten Leber der Crustaceen. *Archv. Mk. Anat.*, vol. xvii., 1880.
37a. WOODWARD, H.—Article "Crustacea," *Enc. Brit.*, 9th Edit.

EARTHWORM.

38. BEDDARD, F. E.—On the Anatomy and Histology of Pleurochæta Moseleyi. *Trans. Royal Soc. Edinburgh*, 1882.
39. BEDDARD, F. E.—Note on Some Earthworms from India. *Ann. N. Hist.*, 1883.
40. BEDDARD, F. E.—On the Anatomy of a Gigantic Earthworm, etc. *Pr. Z. Soc.*, 1884.
41. BLOOMFIELD, J. E. — On the Development of the Spermatozoa of Lumbricus. *Q. J. M. S.*, vol. xx., 1880.
42. CLAPAREDE, E.—Histologische Untersuchungen über den Regenwurm. *Zeit. W. Zool.*, vol. xix., 1869.
43. DARWIN, C.—The Formation of Vegetable Mould through the Action of Worms, etc. 1881.

44. GEGENBAUR, C. — Ueber die sogenannten Respirationsorgane des Regenwurms. *Zeit. W. Zool.*, vol. iv., 1858.
45. HERING, E. — Zur Anatomie und Physiologie der Generationsorgane des Regenwurms. *Zeit. W. Zool.*, vol. viii., 1857.
46. KLEINENBERG, N. — The Development of Lumbricus trapezoides. *Q. J. M. S.*, vol. xix., 1879.
47. KOWALEVSKY, A. — Embryologische Studien an Würmern und Arthropoden. *Mem. Akad. St. P.*, vol. xvi., 1871.
48. LANKESTER, E. R. — The Anatomy of the Earthworm. *Q. J. M. S.*, vol. iv. and v., 1864-5.
49. LANKESTER, E. R. — The Red Vascular Fluid of the Earthworm a Corpusculated Fluid. *Q. J. M. S.*, vol. xviii., 1878.
50. MOJSISOVICS, A. v. — Die Lumbriciden Hypodermis *Stzb. Wien*, vol. lxxvi., 1877.
51. PERRIER, E. — Études sur l'organisation des Lombriciens terrestres. *Arch. Z. Exp.*, vol. ix.
52. PERRIER, E. — Mémoire pour servir à l'histoire des Lombriciens terrestres. *Arch. Mus. Paris*, 1872.
53. POWER, D. — On the Endothelium of the Body Cavity and Blood Vessels of the Common Earthworm, etc. *Q. J. M. S.*, vol. xviii., 1878.
54. ROBINET, CH. — Recherches Physiologiques sur la Sécrétion des glandes de Morren du Lumbricus terrestris. *Cpt. Rend.*, vol. xcvii. 1888.
55. ROLLESTON, G. — The Blood-corpuscles of the Annelids. *Jour. Anat.*, vol. xii., 1878.

SNAIL AND MUSSEL.

56. BARFURTH, D. — Ueber den Bau und die Thätigkeit der Gasteropoden Leber. *Arch. Mk. Anat.*, vol. xxii., 1883.
See also *Preliminary Note, Zool. Anz.*, 1880.
57. BAUDELOT, M. E. — Sur l'Appareil Générateur des Mollusques Gasteropodes. *Paris*, 1863.
58. BLOOMFIELD, J. E. — See No. 1 this list.
59. BÖHRING, L. — Beiträge zur Kenntniss des Centralnervensystems einiger Pulmonaten Gasteropoden. *Leipzig*, 1868.
60. BRAUN, M. — Postembryonale Entwickelung der Süsswasser-Muscheln. *Zoologischer Garten*.
61. BRONN AND KEFERSTEIN. — Klassen und Ordnungen des Thierreichs. Vol. iii. Malacozoa. *Leipzig*, 1862-6.
62. CARRIÈRE, J. — Die Embryonale Byssusdrüse von Anodonta. *Zool. Anz.*, 1884.
63. CATTIE, J. — Ueber die Wasseraufnahme der Lamellibranchiaten. *Zool. Anz.*, 1883.
64. FLEMMING, J. — Ueber die Blutzellen der Acephalen und Bemerkungen ueber deren Blutbahn.
Also — Bemerkung zur Injectionstechnik bei Wirbellosen. *Arch. Mk. Anat.*, vol. xv., 1878.
65. GRIESBACH, H. — Wasseraufnahme bei den Mollusken. *Zool. Anz.*, 1884.
66. JHERING, H. VON. — Ueber die Entwickelung von Helix. *Zeitsch. Jena.*, vol. ix., 1875.
67. JHERING, H. VON. — Zur Kenntniss der Eibildung bei den Muscheln. *Zeit. W. Zool.*, vol. xxix., 1877.
68. JHERING, H. VON. — Zur Morphologie der Niere der sogenannten Mollusken. *Zeit. W. Zool.*, vol. xxix., 1877.
69. JOURDAIN, S. — Sur la conformation de l'Appareil de la Génération de l'Helix aspersa dans le jeune âge. *Rec. Sci.*, 1880.
70. KEBER, G. A. — Beiträge zur Anatomie und Physiologie der Weichthiere. *Königsberg*, 1851.
71. LANGER, K. — Das Gefäss-System der Teichmuschel. *Denk. Kais. Akad.*, 1855-6.
72. LANKESTER, E. R. — On the Development of the Pond Snail. *Q. J. M. S.*, vol. xiv., 1874.
73. LANKESTER, E. R. — Article "Mollusca," *Enc. Brit.*, 9th Edition.
74. LANKESTER, E. R. — The Supposed Taking-in and Shedding-out of Water, in Relation to the Vascular System of Molluscs. *Zool. Anz.*, 1884.
75. MITSUKURI, K. — On the Structure and Significance of some Aberrant Forms of Lamellibranchiate Gills. *Q. J. M. S.*, vol. xxi., 1881.
76. NEUMAU, O. Beiträge zur Anatomie und Physiologie der Pulmonaten. *Tübingen*, 1870.
77. PECK, R. H. — The Minute Structure of the Gills of Lamellibranchiate Mollusca. *Q. J. M. S.*, vol. xvii., 1877.
78. PENROSE, F. G. Note on the Vascular System of Lamellibranchs.
Report of the Committee on the Zool. Station, Naples. *British Association Reports*, 1882.
79. RÜCKER, A. — Über die Bildung der Radula bei Helix pomatia. *Ber. Gies.*, 1888.
80. SOCHACZEWER, D. VON. — Das Riechorgan der Landpulmonaten. *Zeit. W. Zool.*, vol. xxxv., 1880.
81. SPENGEL, J. W. Die Geruchsorgane und das Nervensystem der Mollusken. *Zeit. W. Zool.*, vol. xxxv., 1881.

HYDRA.

82. ALLMAN, G. J.—On the Gymnoblastic Hydroids. *Ray-Society's Publications*, 1871.
83. BRASS, A.—Untersuchungen der Histologie von Hydra viridis. *Zeitschrift für die gesammt. Naturwiss.* vol. liii., 1880.
84. ENGELMANN, T. W.—Ueber Trembley's Umkehrungsversuch an Hydra. *Zool. Anz.*, vol. i., 1878.
85. HAACKE, W. — Zur Speciesunterscheidung in der Gattung Hydra. *Zool. Anz.*, vol. ii., 1879.
86. HAMANN, O.—Zur Entstehung und Entwicklung der grünen Zellen bei Hydra. *Zeit. W. Zool.*, vol. xxxvii., 1882.
87. HARTOG, M. M.—On the Means by which Hydra swallows its Prey. *Proc. Manchester Literary and Philosophic Society*, vol. xix., 1881.
88. HICKSON, S. J.—Abnormal Appearances of Hydra viridis. *Jour. R. M. Soc.*, vol. i., 1878.
89. JICKELI, C. F.— Ueber Hydra. *Zool. Anz.*, vol. v., 1882.
90. JUNG, H.—Beobachtungen über die Entwicklung des Tentakel Kranzes von Hydra. *Morph. Jahrb.*, vol. viii., 1882.
91. KLEINENBERG, N.—Hydra. Eine Anatomisch Entwickelungsgeschichtl. Untersuchung. *Leipzig*, 1872. Abstract by *Allmann in Q. J. M. S.*, vol. xiv., 1874.
92. KOROTNEFF, A. VON.—Zur Kenntniss der Embryologie von Hydra. *Zeit. W. Zool.*, vol. xxxviii., 1883.
93. LANKESTER, E. R.—Article "Hydrozoa," *Enc. Brit.*, 9th Edit.
94. LANKESTER, E. R. On the Chlorophyll-corpuscles and Amyloid Deposits of Spongilla and Hydra. *Q. J. M. S.*, vol. xxii., 1882.
See also a postscript to a paper by F. O. Bower, On the Origin and Morphology of Chlorophyll Corpuscles. *Q. J. M. S.*, vol. xxiv., 1884.
95. LANKESTER, E. R.—On the Intra-cellular Digestion and Endoderm Cells of Lymnocodium. *Q. J. M. S.*, vol. xxi., 1881.
96. MARSHALL, W.—Ueber einige Lebenserscheinungen der Süsswasserpolypen und über eine neue Form von Hydra viridis. *Zeit. W. Zool.*, vol. xxxvii., 1882.
97. PARKER, T. J. On the Histology of Hydra fusca. *Pr. R. Soc.*, 1880. *Also Q. J. M. S.*, vol. xx., 1880.

UNICELLULAR ORGANISMS.

98. AUERBACH, L.—Ueber die Einzelligkeit der Amœben. *Zeit. W. Zool.*, vol. vii., 1856.
99. BREFELD, O. Botanische Untersuchungen über Schimmelpilze. Part iv., *Leipzig*, 1881.
100. BUTSCHLI, O.—Bronn's Klassen und Ordnungen des Thierreichs. Vol. i., *Protozoa, Leipzig*, 1880-4.
101. COHN. -Untersuchungen über Bacterien. *Beiträge zur Biologie der Pflanzen, Breslau. Heft 2,* 1872. *Heft* 3, 1875. *Band* 2, *Heft* 2, 1876.
See also *Q. J. M. S.*, vol. xiii., 1873; vol. xvi., and vol. xvii., 1876-7.
102. DALLINGER, W. H.—On the Measurement of the Diameter of the Flagella of Bacterium termo, etc. *Jour. R. M. Soc.*, vol. i., 1878.
103. ENGELMANN, T. W.—Zur Physiologie der Contractilen Vacuolen der Infusionsthiere. *Zool. Anz.*, vol. i., 1878.
See also a *Note in the same*—Ueber Gasentwicklung im Protoplasma lebender Protozoen.
104. EWART, J. C. On the Life-History of Bacillus anthracis. *Q. J. M. S.*, vol. xviii., 1878.
105. EWART, J. C.—On the Life-History of Bacterium termo and Micrococcus. *Pr. R. Soc.*, 1878.
106. GEDDES AND EWART. - On the Life-History of Spirillum. *Pr. R. Soc.*, 1878.
107. GRUBER, A.- Beiträge zur Kenntniss der Amöben. *Zeit. W. Zool.*, vol. xxxvi., 1882.
108. GRUBER, A. — Ueber Kerntheilungsvorgänge bei einigen Protozoen. *Zeit. W. Zool.*, vol. xxxviii., 1883.
109. GRUBER, A.— Ueber Kern und Kerntheilung bei den Protozoen. *Zeit. W. Zool.*, vol. xl., 1884.
110. GRUBER, A.—Studien über Amöben. *Zeit. W. Zool.*, vol. xli., 1884.
111. HUXLEY, T. H.—Article on "Yeast," *Contemporary Review*, 1871.
112. JICKELI, C. F. -Ueber die Kernverhältnisse der Infusorien. *Zool. Anz.*, vol. vii., 1884.
113. KENT, W. S. -A Manual of the Infusoria. *London*, 1880-81.
114. KLEBS, G.—Beiträge zur Kenntniss niederer Algenformen. *Bot. Zeit.*, 1881.
115. KLEIN, E. -Koch's New Method of Pure Cultivation of Bacteria. *Q. J. M. S.*, vol. xxi., 1881.
116. LANKESTER, E. R.—On Bacterium rubescens, etc. *Q. J. M. S.*, vol. xvi., 1876.

117. LANKESTER, E. R.—On Lithamœba discus. Q. J. M. S., vol. xix., 1879.
118. LEIDY, J.—Freshwater Rhizopods of North America. *Publications of the United States Geographical Survey, Washington*, 1879.
119. NÄGELI, C. VON.—Theorie der Gärung. *München*, 1879.
120. REES MAX.—Botanische Untersuchungen über die Alkoholgährungepilze. *Leipzig*, 1870.
See also A. W. Bennett. Some Account of Modern Researches into the Nature of Yeast. Q. J. M. S., vol. xv., 1875.
121. RYDER, J. A. — On the Chlorophyll Granules in Vorticella.
Proc. United States National Museum, vol. vii., 1884.
122. SALLITT, JESSIE. — On the Chlorophyll Corpuscles of some Infusoria. Q. J. M. S., vol. xxiv., 1884.

122A. SCHÜTZENBERGER, P.—On Fermentation. *Internat. Scientific Series*. 1876.
123. TYNDALL, J. — The Optical Deportment of the Atmosphere, in Relation to the Phenomena of Putrefaction and Infection. *Phil. Trans.*, 1876.
124. TYNDALL, J.—Further Researches on the Deportment and Vital Persistence of Putrefactive and Infective Organisms from a Physical Point of View. *Phil. Trans.*, 1877. See also the same author's " Floating Matter of the Air, etc." *London*, 1881.
125. WALLICH, G. C.—On the Value of the Distinctive Characters of Amœba. *Ann. N. Hist.*, 1868.
See also other papers in the above, by the same author.

FUNGI.

126. DE BARY, A., AND WORONIN.—Beiträge zur Morphologie und Physiologie der Pilze. Vol. i., *Frankfort*, 1864-70.
127. DE BARY, A.— Vergleichende Morphologie und Biologie der Pilze, etc. *Leipzig*, 1884.
128. BREFELD, O. — Botanische Untersuchungen der Schimmelpilze. Vol. i., *Zygomycetes*. *Leipzig*, 1872.

129. BREFELD, O. — The Same. Vol. ii., *Penicillium*. *Leipzig*, 1874.
130. CUNNINGHAM.—Conidial Fructification in Mucorini. *Trans. Linnæan Soc., Botany*, 1879. .
131. VAN TIEGHEM ET LE MONNIER.—Recherches sur les Mucorinées. *Ann. Sci. Nat. Botanique*, vol. xvii., 1873.
Abstract by Dyer, Q. J. M. S., vol. xiv., 1874.

GREEN PLANTS.

132. DE BARY, A.—Vergleichende Anatomie der Vegetationsorgane der Phanerogamen und Farne. *Leipzig*, 1877.
English Translation by Bower and Scott, 1885.
133. GARDINER, W. — Middle Layer of the Cell Wall. *Proc. Cambridge Philosophical Society*, vol. v., 1884.
134. GROVES, H. AND J. — British Characeæ. *Trimen's Journal of Botany*, vol. ix., 1880.
135. JOUOW, F.—Die Zellenkerne von Chara fœtida. *Bot. Zeit.*, vol. xxix., 1881.
136. LUERSSEN, C.—Medicinisch-Pharmaceutische Botanik, etc. *Leipzig*, 1882.
137. PARKER, T. J.—On some Applications of Osmic Acid to Microscopic Preparations. *Jour. R. M. Soc.* 1879.

138. RUSSOW, E. — Vergleichende Untersuchungen der Leitb ndel-Kryptogamen, etc. *St. Petersburg und Leipzig*, 1872.
139. SCHENCK, A. — Handbuch der Botanik. *Breslau*, 1881-4.
140. STRASBURGER, E.—Ueber Befruchtung und Zelltheilung. *Leipzig*, 1878.
141. STRASBURGER, E. — Die Angiospermen und die Gymnospermen. *Jena*, 1879.
142. STRASBURGER, E. — Ueber Zellbildung und Zelltheilung. *Leipzig*, 1880.
143. STRASBURGER, E.—Neue Untersuchungen über den Befruchtungsvorgang bei den Phanerogamen, etc. *Jena*, 1884.

www.ingramcontent.com/pod-product-compliance
Lightning Source LLC
Chambersburg PA
CBHW020302170426
43202CB00008B/472